WASTEWATER TREATMENT USING GENETICALLY ENGINEERED MICROORGANISMS

HOW TO ORDER THIS BOOK

BY PHONE: 800-233-9936 or 717-291-5609, 8AM–5PM Eastern Time

BY FAX: 717-295-4538

BY MAIL: Order Department
Technomic Publishing Company, Inc.
851 New Holland Avenue, Box 3535
Lancaster, PA 17604, U.S.A.

BY CREDIT CARD: American Express, VISA, MasterCard

WASTEWATER TREATMENT USING GENETICALLY ENGINEERED MICROORGANISMS

DR. MASANORI FUJITA
Professor
Department of Environmental Engineering
Osaka University, Japan

DR. MICHIHIKO IKE
Research Associate
Department of Environmental Engineering
Osaka University, Japan

LANCASTER · BASEL

Wastewater Treatment Using Genetically Engineered Microorganisms
a **TECHNOMIC**® publication

Published in the Western Hemisphere by
Technomic Publishing Company, Inc.
851 New Holland Avenue, Box 3535
Lancaster, Pennsylvania 17604 U.S.A.

Distributed in the Rest of the World by
Technomic Publishing AG
Missionsstrasse 44
CH-4055 Basel, Switzerland

Printed in the United States of America
10 9 8 7 6 5 4 3 2 1

Main entry under title:
Wastewater Treatment Using Genetically Engineered Microorganisms

A Technomic Publishing Company book
Bibliography: p.

Library of Congress Catalog Card No. 94-60490
ISBN No. 1-56676-139-5

TABLE OF CONTENTS

PREFACE

Water quality management is one of the world's most significant social necessities. As industry and society become more complex, water pollution problems increase. A growing number of man-made products cause serious public health problems because of their toxicity and low biodegradability. Clean, safe water is necessary both for everyday life and for industry. Therefore, more efficient and advanced wastewater treatment techniques are required immediately.

In recent years, genetic engineering has been proposed by several researchers as a way to improve wastewater treatment. The potential for enhanced breakdown of hazardous organic compounds, such as aromatic, haloaromatic, and haloaliphatic compounds, synthetic polymers, pesticides, and so on, by using genetically engineered microorganisms (GEMs) having high degradation activities is especially promising. However, this approach is hardly beyond the planning stage because of the lack of basic research into the ecology, physiology, and genetics of wastewater treatment processes and GEMs. Also, there has been little practical research into the breeding and characterization of GEMs for wastewater treatment, studies on the survival and gene expression of GEMs involved in the process, and risk assessments of biohazard.

We have been researching the application of GEMs to wastewater treatment, especially in the activated sludge process, for a decade. This book, which summarizes and systematizes the results of this research, as well as other research reported in the literature, aims to suggest a strategy and guidelines on actual uses of GEMs in wastewater treatment.

Chapter 1 describes the basic concept, including an introduction to biological wastewater treatment and genetic engineering. Research on bacterial populations and their plasmids of wastewater treatment processes are contained in Chapters 2 and 3, which contain basic ecological and genetic information. Chapters 4 and 5 show the benefits of genetic engineering applications by introducing some case studies on the breeding of wastewater treatment bacteria. Chapters 6 and 7 deal with the genetic and ecological stability of GEMs, which are most important problems. In Chapter 8, a few forms of application of GEMs to wastewater treatment are proposed and evaluated. Chapter 9 concludes the book with a discussion on the future of this technique.

CHAPTER 1

Basic Concepts

Biological wastewater treatment began about a century ago, and is considered a well-established technology. However, water quality requirements have become more stringent in recent years, as more pollutants are recognized, and as the public places more importance on water quality. The time seems ripe for a new concept in wastewater treatment—the application of genetic engineering and biotechnology to wastewater treatment microorganisms.

Chapter 1 describes the basic concept of this new approach. First, recent problems concerning biological wastewater treatment are reviewed and explained from the microbial point of view. Second, a brief introduction to genetic engineering is given so as to clarify the availability of this approach. Last, the strategy of the breeding and application of genetically engineered microorganisms (GEMs) is discussed.

1.1 PROBLEMS OF BIOLOGICAL WASTEWATER TREATMENT

Biological wastewater treatment depends on the metabolic activity and physical characteristics of the microorganisms involved in the process. Organic compounds—nutrients such as nitrogen and phosphorous—and some minerals in the wastewater are removed by the metabolism of various bacterial populations, and consequently their cells are synthesized by anabolism. Protozoa are predatory, feeding on bacteria in most cases. Fungi, other procaryotes, viruses, and so on also are part of the treatment processes. Together, they form complex mixed microbial communities of cultures supported in a wastewater medium. In the activated sludge process, the settling characteristics of the culture significantly affect its treatment efficiency.

Therefore, a most important concern in wastewater treatment is to develop and maintain suitable cultures of mixed microbes.

Conventionally, the improvement of biological wastewater treatment processes has been carried out by controlling environmental conditions so as to proliferate and maintain desirable microorganisms and their activities, and/or to eliminate undesirable microorganisms and their activities (indirect control of microorganisms). The significant control parameters are pH, dissolved oxygen concentration (DO), oxidation-reduction potential (ORP), load of organic compounds and nutrients, water flow, and so on.

A practical example of the conventional approach is the treatment process for nitrogen removal through sequential nitrification and denitrification, in which nitrifying bacteria (*Nitrosomonas* sp. and *Nitrobacter* sp.) and denitrifying bacteria are kept under optimal conditions separately (Barnard, 1970). Selecting for the prevention of filamentous bulking in activated sludge processes is another example, in which growth of filamentous bacteria such as *Sphaerotilus* sp. are suppressed by controlling the substrate concentration (Chudoba et al., 1973; 1974). Physical removal of *Nocardia* sp. was proposed as a countermeasure for the problems of bacterial foaming and scum formation in activated sludge processes (Richards et al., 1990).

Although various wastewater treatment processes have been developed, a number of common problems cannot be solved completely by conventional approaches, suggesting some technical limits to the usual techniques. These problems include:

- Xenobiotic and recalcitrant compounds (aromatics, haloaromatics, haloaliphatics, polycyclic aromatics, pesticides, etc.) cannot be degraded completely or efficiently.
- Removal of nutrients (nitrogen and phosphorous) is not sufficient.
- Influx of toxic compounds (cyanide, heavy metals, antibiotic agents, etc.) lowers treatment efficiency.
- Treatment under extraordinary conditions (extremely low or high pH and/or temperature, high salinity, etc.) is impossible.
- Constant control is difficult because of the fluctuation of environmental conditions (quantity and quality of influent, atmospheric temperature, etc.).
- The problems of bulking caused by filamentous bacteria, and foaming caused by actinomycetes, are often observed in the activated sludge process.

These problems seem to involve the microorganisms themselves. Desirable microorganisms for efficient treatment are usually present in relatively small numbers, and their physiological characteristics cannot be manipulated by controlling environmental factors. For example, if no degrading microorganisms are contained in the treatment processes, xenobiotic compounds will not break down. Even when degrading microorganisms are present, they aren't as effective as they could be because their growth rates are lower than those of the other species present in the treatment environment. Fluctuating environmental conditions influence bacterial metabolism, rates of nitrification catalyzed by *Nitrosomonas* sp. and *Nitrobacter* sp., and phosphorous accumulation by bacteria. Microorganisms without resistance are killed by the influx of toxicants. It is evident that there is a need to improve the natural functions of wastewater treatment microorganisms.

1.2 APPLICATION OF GENETIC ENGINEERING

Genetic engineering—molecular breeding—began in the 1960s as a technique to alter the characteristics of microorganisms and other animals and plants to create man-made mutants through manipulation of the genetic information coded on the desoxyribonucleic acid (DNA) in cells. With genetic engineering, specialists can, in theory, create organisms with desirable characteristics.

Numerous genetically engineered organisms have been constructed for a wide variety of applications. The drug industry uses GEMs to produce human insulin, hormones, and other medical agents. The food industry also uses various GEMs. Agriculture and forestry researchers use GEMs in plant breeding and to produce microbial pesticides.

GEM technology offers a new and attractive approach for improving biological wastewater treatment. If GEMs with desirable and controllable characteristics are constructed and fully utilized, the performance and efficiency of treatment processes can be greatly enhanced. The special promise of this approach for the removal of hazardous compounds has been emphasized by some researchers, although the promise has not yet been realized (Kobayashi, 1984; McClure et al., 1991; Wyatt, 1988). This approach can be regarded as a specialized form of *bioremediation,* which means the application of indigenous and/or cultivated (inoculated or introduced) organisms—including genetically engineered organisms—to polluted hazardous waste sites in order to accelerate environmental restoration.

In theory, GEMs can do anything in wastewater treatment. For example, fast-growing GEMs with high degradation activity may enhance the breakdown of xenobiotic compounds. Nitrifying bacteria, denitrifying bacteria and phosphorous-accumulating bacteria can be also genetically modified to improve and stabilize nutrient removal. If genes encoding toxin reduction and resistance are introduced into sludge microorganisms, performance of

TABLE 1.1. Presumable targets for possible GEM-applications.

Bio-Transformation of Pollutants
Improvement of degradation activity (xenobiotics)
Extention of catabolic range
Enhanced nutrient removal
Resistance to toxic compounds
Degradation under high/low pH and temperature
Solid / Liquid Separation
Improvement of flocculation and biofilm attachment
Suppression of filamentous/foaming growth
General Treatment Performance
Enhanced BOD and COD reductions
Process stabilization (resistance to shock-loading)

the treatment may be little affected by shock loading of toxicants. By analyzing and manipulating genes associated with bacterial bulking, foaming, flocculation, and biofilm sloughing, researchers may be able to develop troublefree treatment processes. Ultimately, genetic engineering may be able to improve wastewater treatment processes in general, such as BOD and COD removal. Potential GEM applications to wastewater treatment are summarized in Table 1.1.

1.3. STRATEGY

Feasibility is another major concern. It is difficult to breed GEMs which possess only desirable characteristics. Introduced or manipulated genetic information in the GEMs normally occupies a minor part of a cell's total genome. Unstable GEM genotypes and phenotypes are another problem. Both failures and successes for microbial inoculation have been reported, but the reasons for success or failure are often not known (Goldstein et al., 1985). Therefore, research into microbial ecology, including the survival of GEMs in the wastewater treatment process, is considered necessary to establish methods for the successful use of GEMs. Details of these technical problems will be described and discussed in the following chapters.

A strategy for applying GEMs to wastewater treatment, and which takes these technical problems into consideration, needs to be worked out. One possible strategy is shown in the schematic flow chart in Figure 1.1.

1.3.1 Identifying the Problem

First, wastewater treatment problems which cannot be solved by existing processes/systems and their microbial aspects—the relation of causes and effects—are surveyed, and the desirable GEMs or microbial characters are presumed present. Examples of conceptual GEMs and genetic manipulations necessary for the solution of several problems contained in Table 1.1 were already described in Section 1.2. Actual GEMs are created step-by-step with close adherence to legal regulations and guidelines for field applications of GEMs. Public opinion must also be taken into account.

1.3.2 Selecting Materials and Techniques

Second, materials for breeding the proposed GEMs must be selected, and suitable breeding procedures and techniques determined. Required materials include the objective genes and ordinary host microorganisms. Various genes and microorganisms which seem to be useful, and which have already been isolated and described by many researchers, are shown in

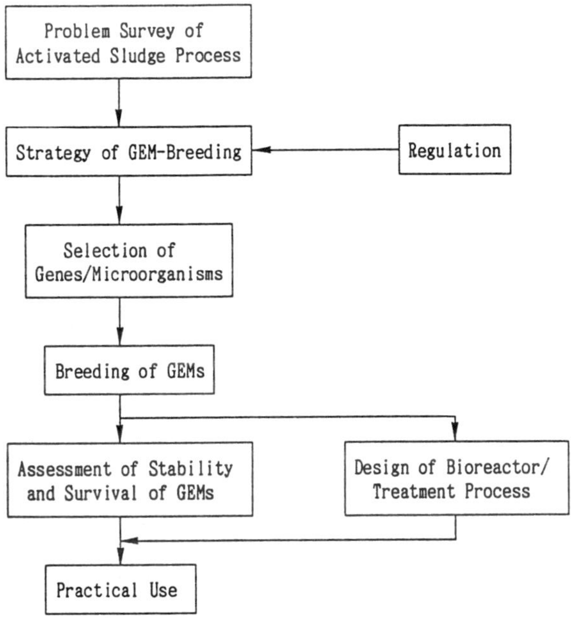

FIGURE 1.1. Schematic flow of application of GEMs to the wastewater treatment.

TABLE 1.2. Xenobiotic-degrading microorganisms.

Compound	Reported Species or Cultures
Halogenated Aliphatics	
Trichloroethane	Marine bacteria
Trichloromethane	Sewage sludge
Trichloroethylene	Soil bacteria, methanogenic culture
Aromatics	
Benzene	*Pseudomonas putida*
Toluene	*Bacillus* sp., *P. putida*
Cresol	*Pseudomonas*
Phenol	*Pseudomonas, Vibrio, Flavobacterium,* etc.
Nitrophenol	Rumen microorganisms
Pentachlorophenol	Soil bacteria
Polycyclic Aromatics	
Naphthalene	*Pseudomonas, Alcaligenes, Flavobacterium,* etc.
Biphenyl	*Pseudomonas putida*
Pyrene	Stabilization pond organisms
Pesticides	
Dieldrin	*Pseudomonas,* Actinomycetes
DDT	Soil bacteria, yeasts, sewage, *Klebsiella*

*Referring to Kobayashi and Rittmann (Kobayashi and Rittmann, 1982).

Table 1.2, which gives examples of microorganisms which degrade xenobiotic compounds.

Genetic manipulation techniques are limited by the sources and types of genes, species of host microorganisms, and their combinations. A wide variety of techniques as described in such manuals as *Molecular Cloning* (Maniatis et al., 1982) may often provide valuable aids in the breeding of GEMs. When existing techniques cannot be applied to selected materials, new or modified methods must be developed. Design of desirable GEMs are mostly determined in this stage.

1.3.3 Feasibility Assessment

Third, the feasibility of improving wastewater treatment using the constructed GEMs is assessed by investigating their functional abilities, especially concerning the expression of desirable genes, genetic stability, and the survival and fate in the processes. In this stage, experimental researches are more or less required. Laboratory tests, which can be carried out under several different conditions, are useful for the characterization and screening of GEMs. From lab results, the potentials of the constructed GEMs are evaluated and the most desirable selected. The pilot tests, which simulate actual treatment processes, should be performed afterwards for the selection, development, synthesis, and optimization of bioreactors and/or treatment processes using selected GEMs.

1.3.4 Assessing Risks and Benefits

Finally, the risks and benefits of a newly developed process using GEMs are compared for performance with existing processes before a decision on whether or not to use the GEM is made.

1.4 REFERENCES

Barnard, J. L. 1970. "Biological Denitrification," *Wat. Pollut. Cont.*, 72:705–720.

Chudoba, J., P. Grau and V. Madera. 1973. "Control of Activated Sludge Bulking, II. Selection of Microorganisms by Means of a Selector," *Wat. Res.*, 7:1389–1406.

Chudoba, J., J. Blaha and V. Madera. 1974. "Control of Activated Sludge Bulking, III. Effect of Sludge Loading," *Wat. Res.*, 8:231–237.

Goldstein, R. M., L. M. Mallory and M. Alexander. 1985. "Reasons for Possible Failure of Inoculation to Enhance Biodegradation," *Appl. Environ. Microbiol.*, 50:977–983.

Kobayashi, H. A. 1984. "Application of Genetically Engineering to Industrial Waste/Wastewater Treatment," in *Genetic Control of Environmental Pollutants*, G. S. Omenn and A. Hollaender, eds., New York-London: Prenum Press, pp. 47–80.

Kobayashi, H. and B. E. Rittman. 1982. "Microbial Removal of Hazardous Compounds," *Environ. Sci. Technol.*, 16:170–183.

Maniatis, T., E. F. Fritsch and J. Sambrook. 1982. *Molecular Cloning—A Laboratory Manual.* Cold Spring Harbor, New York: Cold Spring Harbor Laboratory.

McClure, N. C., J. C. Fry and A. J. Weightman. 1991. "Genetic Engineering for Wastewater Treatment," *J. IWEM.*, 5:608–616.

Richards, T., P. Nungesser and C. Jones. 1990. "Solution of *Nocardia* Foaming Problems," *Res. J. Wat. Pollut. Cont. Fed.*, 62:915–919.

Wyatt, J. M. 1988. "Biotechnological Treatment of Industrial Wastewater," *Microbiol. Sci.*, 5:186–190.

CHAPTER 2

Bacterial Ecology of Wastewater Treatment Processes

A significant reason for the delay in GEM applications is the lack of basic knowledge of the microbial ecology and genetics of wastewater treatment processes. Research on these basic subjects and the practical uses of GEMs will help assess the behaviors and fates of GEMs or recombinant DNAs in actual treatment processes. This chapter will deal with the bacterial ecology of wastewater treatment processes, mainly the activated sludge process which is a typical aerobic process. Although there has been much research into activated sludge microbiology, not much useful knowledge has been obtained on the application of GEMs.

After an introduction to the bacteriology of various wastewater treatment processes, this chapter presents a study on the relationship between an important operating parameter—SRT (sludge retention time)—and bacterial flora of the activated sludge process. The research was designed to obtain results which could be generalized.

2.1 ROLES OF BACTERIAL POPULATION

2.1.1 Niches

The microbial community of the wastewater treatment process is specialized, but includes a complex mix of viruses, procaryotic organisms (bacteria, etc.), single-celled eucaryotic organisms (protozoa, fungi, yeasts, etc.), and simple invertebrates (rotifers, nematodes, aquatic insects, insect larvae, crustaceans, etc.).

Among these, bacteria are the most important because they occupy the most basic food niche, and they can utilize a wide variety of organic compounds directly from the wastewater. Also, bacterial numbers are much greater than the other microorganisms. In a sense, the functions and properties of the wastewater treatment processes are determined by the abilities and physiological characteristics of the bacterial population.

Protozoa, which do not participate directly in the breakdown of pollutants, are also considered important as predators of bacteria. On the other hand, fungi and yeasts appear rarely in the treatment processes.

2.1.2 General Roles

The general roles of bacterial populations in wastewater treatment are the breakdown of organic compounds and the uptake of nutrients, processes that remove organic (BOD, COD) and some mineral pollutants. These activities depend on normal bacterial metabolism—catabolism of carbohydrates, fats, protein, etc., and assimilation of cell molecules from carbon, nitrogen, and phosphorous sources, and several other minerals and nutrients. In both aerobic treatment (activated sludge process, trickling filter, etc.) and anaerobic treatment (anaerobic digestion, methane fermentation), either harmless (CO_2, H_2O) or easily removable end products (cells, CH_4) are produced. The typical (and simplified) biochemical processes in wastewater treatment are expressed as follows.

In the aerobic processes:

$$\text{organic compounds} + O_2 \rightarrow \text{oxidized organic compounds} + \text{cells} + CO_2 + H_2O \tag{2.1}$$

In the anaerobic processes:

$$\text{organic compounds} \rightarrow \text{hydrolyzed organic compounds} + \text{cells} + CO_2 + H_2O + CH_4 \tag{2.2}$$

2.1.3 Specific Roles

Specific roles of bacterial populations in the wastewater treatment processes are listed in Table 2.1. These roles are generally associated with specific groups of bacteria which may not be dominant in the processes. For example, nitrification is mediated by limited species—ammonium-oxidizing bacteria (*Nitrosomonas* sp., etc.) and nitrite-oxidizing bacteria (*Nitrobacter* sp., etc.), whereas denitrification is carried out by a wide variety of bacterial populations. In research by Pike and Carrington (1972),

TABLE 2.1. *Specific roles of wastewater treatment bacteria.*

Nitrification
Denitrification
Phosphate accumulation
Xenobiotic degradation
Detoxification (heavy metals, toxicants)
Floc formation
Attached growth

Nitrosomonas sp. counts are approximately 10^{-5} times less than the total bacterial count. In spite of the relatively small population, these specific bacteria are considered important, because of their useful abilities in wastewater treatment. Accordingly, major targets of the application of genetic engineering are their specific abilities.

2.2 BACTERIAL FLORA

2.2.1 Selective Mechanisms

Bacterial flora form according to selective pressures operating within the treatment processes. The fraction of a bacterial population is determined by its growth and disappearance, which depends on the decay and losses in effluent and biomass wastage. In principle, the bacterial population which grows the fastest and decays the slowest within a particular treatment process will dominate. The growth and/or decay rates of bacteria are considerably influenced by two major selective factors: the environmental conditions and the ecological relationships between the organisms present.

The significant environmental factors are nutrition (variety and concentrations of available substrates, nutrient limitation, etc.), oxygen supply (DO), temperature, pH, and inhibitors and/or toxicants. Although the effects of these factors upon the growth of certain bacteria can be determined in pure cultures, those in a mixed community, such as the wastewater treatment process, cannot be easily assessed because of the complex ecological interactions.

Various forms of microbial interactions are listed in Table 2.2. The interactions shown in the table are defined only between two members. In fact, numerous members participate in intricate ways in the microbial community of a wastewater treatment process. Among these interactions, com-

*TABLE 2.2. Various forms of microbial interactions.**

Interactions	Definition
Nutralism	Lack of interaction
Competition	A race for nutrients, space, etc.
Mutalism	Each member benefits from the other
Commensalism	One member benefits while the other unaffected
Amensalism	One member adversely affects the other
Parasitism	One organism steals from the other
Predation	One organism ingests the other

*Harder (1981).

petition is considered to play the most significant role in the selection of bacterial populations and the formation of bacterial flora.

As the substrate grows, competition among various members of the microbial community can be observed, since there is a variety of bacteria on each trophic level. Further, the organic compounds contained in the influent will be oxidized and/or hydrolyzed into their metabolites, causing secondary competition between the members capable of utilizing the metabolites. Nitrogen, phosphorous and oxygen in the aerobic processes are other materials for which there may exist competition among the bacterial members. Other nutrients such as sulfur, potassium, magnesium, and the trace elements may not cause competition because they exist in sufficient amounts in the process.

Mutalism, including commensalism, also seem to occur often, and considerably change bacterial flora. These interactions, observed in studies on the microbial degradation of organic compounds—especially the xenobiotic compounds—are seen in the following examples (Harder, 1981). Interactions in wastewater treatment communities have not been elucidated clearly.

- A two-member culture growing on glucose consisted of the strains of phenylalanine-requiring *Lactobacillus plantarum* and folic acid-requiring *Streptococcus faecalis,* neither of which can grow in individual pure cultures. The former provided folic acid, and the latter reciprocated with the production of phenylalanine (interdependence of growth factor).
- Bacterial cultures growing on linear alkylbenzene sulphonate (LAS) as the limiting carbon source were dominated by species of *Pseudomonas* and *Alcaligenes,* while none of the isolates alone was able to utilize LAS, suggesting the combined metabolic attack.
- In a three-member mixed culture growing on orcinol as the sole

carbon source, the relief of substrate inhibition was observed. The typical substrate-inhibited growth in the pure culture of *Pseudomonas stutzeri* on orcinol was changed considerably by the presence of *Brevibacterium linens* and *Curtobacterium* sp., which can grow on the metabolites and products of cell lysis provided by *P. stutzeri,* and which may also reduce the inhibitory effects of these products.
- In a community growing on cyclohexane as the carbon source, *Nocardia* sp. was tentatively identified as the primary degrader, and an unidentified pseudomonad grew on its excretion products, which also provided growth factors.

Predation and parasitism are observed mainly between bacteria and protozoa, and between bacteria and viruses, in wastewater treatment communities.

2.2.2 Dominant Bacteria

Well-adapted or selected bacteria under a certain treatment condition may become dominant in the treatment process. There are several methodological difficulties in the quantitative characterization of bacterial flora in wastewater treatment processes. Predominant species cannot be identified, but the bacterial genera, isolated frequently from the aerobic and anaerobic processes by many researchers, and considered dominant, are summarized in Table 2.3 and Table 2.4.

Among the aerobic processes, the bacterial flora of activated sludge have been most studied (Pike, 1975; Dias and Bhat, 1964; Van Gils, 1964). Dominant bacteria in the activated sludge processes belong to genuses *Pseudomonas, Alcaligenes, Flavobacterium,* and other gram negative

TABLE 2.3. Principal bacterial genera in the aerobic processes.

Achromobacter	*Sphaerotilus*
Alcaligenes	*Spirillum*
Azotobacter	*Zoogloea*
Bacterium	*Arthrobacter*
Chromobacterium	*Bacillus*
Comamonas	Coryneform group
Enterobacteriaceae	*Micrococcus*
Flavobacterium	*Mycobacterium*
Lophomonas	*Nocardia*
Nitrosomonas	*Sarcina*
Pseudomonas	*Staphylococcus*

TABLE 2.4. Principal bacterial genera in the anaerobic processes.

Clostridium	*Methanospirillum*
Desulfovibrio	*Methanothrix*
Eubacterium	*Peptococcus*
Methanobacterium	*Propionibacterium*
Methanobrevibacterium	*Syntrophobacter*
Methanosarcina	*Syntrophomonas*

genera such as *Acinetobacter, Zoogloea,* etc. Gram positive bacteria belonging to genuses *Bacillus, Arthrobacter,* and *Micrococcus* have sometimes been isolated as dominants. The bacteriology of other aerobic processes is not so clear. However, activated sludge bacterial flora, ordinarily maintained strictly under aerobic conditions, are considered more specialized and show a lower diversity than other processes, such as trickling filters, which have an anaerobic or partly microaerobic environment in addition to an aerobic part.

On the other hand, more specialized bacterial flora seem to be formed in anaerobic treatment. Almost all bacterial members may involve the two main biochemical processes—the hydrolysis of complex organic compounds into fatty acids and alcohols (primarily butyric, propionic, and acetic acids), and methane fermentation from these hydrolyzed compounds, CO_2, and H_2. The former processes are mediated mainly by *Clostridium, Eubacterium, Peptococcus,* etc. and the latter by *Methanobacterium, Methanothrix, Methanosarcina,* etc.

2.3 CHANGES OF BACTERIAL FLORA IN THE ACTIVATED SLUDGE PROCESS

2.3.1 Process Operation

This section describes changes to bacterial flora in the activated sludge process as a function of SRT (sludge retention time), which is the most significant operational parameter. Also considered are organic loading, hydraulic retention time, the rate of return sludge, etc. SRT is the mean cell residence time, a general word for biological processes, and is defined as follows:

$$\text{SRT} = \text{mass of cells (or activated sludge) in reactor (or aeration tank)} / \text{mass of cells wasted per day} \quad (2.3)$$

Under steady state conditions, sludge wastage equals sludge growth. SRT also means the inverse of the mean specific growth rate (μ) of activated sludge, which is defined as follows:

$$\mu = \text{growth of cells per day/mass of cells in reactor} = \text{SRT}^{-1} \quad (2.4)$$

The model activated sludge unit used in this study consisted of an aeration tank (working volume: V = 6 L) equipped with a filter-separator instead of a sedimentation tank for solid/liquid separation, a setup which made sludge return unnecessary. Synthetic wastewater containing meat extract, peptone, and minerals (total organic carbon concentration: TOC = ca. 160 mg/L) was fed continuously at a flow rate of 5 L/day to the aeration tank. TOC loading for the reactor was approximately 1.6 g-TOC/L/day. Aeration was carried out with a diffuser from the bottom at 0.6 vvm (normal volume of air/volume of reactor/min). Treatment temperature was maintained at 25°C. Since the suspended solid concentration (SS) of the effluent was extremely low, the SRT could be regulated by controlling the rate of sludge wastage (Q_w) withdrawn from the aeration tank, which was expressed as follows:

$$\text{SRT} = V/Q_w \text{ or } Q_w = V/\text{SRT} \quad (2.5)$$

For example, if the SRT was maintained at 10 days, the sludge wastage per day was 10% of the volume of the aeration tank (=0.6 L).

To separate SRT effects on bacterial flora as much as possible from the effects of other factors, seed sludge for the units operated at different SRTs was obtained from a fill-and-draw treatment unit which had been operated under constant conditions for more than 2 years.

2.3.2 Isolation of Bacterial Strains

The continuous activated sludge treatment units were operated at SRTs of 2, 5, 10, and 15 days. TOC removal was approximately 95% at each SRT. However, nitrification and the physical characteristics (color, flocculation, etc.) of the activated sludge changed considerably as SRT varied. Nitrification did not occur at all at an SRT of 2 days, whereas the removal of the ammonium and organic nitrogen, which depends mainly on nitrification, were maintained at relatively high levels (80–90%) at SRTs between 5 and 15 days. Settling characteristics of the sludge were better at SRTs of 5 and 15 days than at 2 days, suggesting that the bacterial flora at the four SRTs were different from each other.

With steady state conditions for each SRT, activated sludge was sampled from each aeration tank as waste sludge, and plated onto CGY agar to nonselectively isolate all bacterial strains that were present in high numbers. Although it is clear that no single medium, especially a synthetic medium, can support the growth of the wide range of bacteria present in activated sludge, CGY agar, composed of 5 g of casitone, 5 g of glycerol, 1 g of yeast extract, and 1 L of water, is generally considered a most suitable medium for the enumeration and isolation of activated sludge bacteria (Pike et al., 1972). The temperature of incubation is also important, and we selected 22°C, which is a little lower than the 25°C used in the preparation of the activated sludge samples.

Another technical difficulty in isolating the bacterial strains is that sludge flocs must be disrupted to give discrete viable cells which are able to form colonies on the medium. In this study, floc dispersal was carried out as follows:

The activated sludge samples were pelletted by centrifugation and resuspended in the same volume of 5 mg/L sodium tripolyphosphate solution, which can prevent reflocculation of bacterial cells once they are released. A 50 mL of sample diluted 10-fold with the solution was treated by sonicator (GT200: Nihon-Seiki Co., Chiyoda-Ku, Tokyo, Japan) at 200 μA for 2.5 min, and then used for the bacterial isolation. Viable bacterial counts of the activated sludge samples obtained in this plating are summarized in Table 2.5.

The viable bacterial counts per mL of mixed liquor increased as SS in the aeration tank (MLSS) increased with the increase in SRT. On the other hand, the counts per g of MLSS (dry weight of mixed liquor suspended solid) did not vary much at SRTs of 5–15 days. At 2 days SRT, MLSS was

TABLE 2.5. Viable bacterial counts at different SRTs.

Activated Sludge	Viable Count	
	cfu/mL	cfu/g-MLSS
Sludge from Laboratory Units		
SRT = 2 days	4.9×10^8	2.0×10^{12}
SRT = 5 days	6.6×10^8	7.6×10^{11}
SRT = 10 days	8.2×10^8	9.7×10^{11}
SRT = 15 days	1.1×10^9	8.6×10^{11}
Sludge from Actual Plants		
T plant	4.6×10^8	5.9×10^{10}
S plant	6.8×10^8	7.2×10^{11}

approximately 2–3 times higher than at longer SRTs, suggesting high viability or metabolic activity.

Almost all colonies formed on each plate were picked up and isolated as the representative strains present in the activated sludge process under a certain SRT. There were 36, 60, 68, and 90 bacterial strains isolated at SRTs of 2, 5, 10, and 15 days, respectively. Additionally, 32 and 30 strains were isolated from activated sludge samples collected at two actual wastewater treatment plants. One is T plant and the other is S plant, where mainly domestic wastewater was treated and SRTs were maintained at more than 30 days and approximately 5 days, respectively.

2.3.3 SRT and Bacterial Flora

In all, 316 strains of bacteria were isolated and their basic physiological characteristics investigated as shown in Table 2.6. Eight identification tests listed in the table classified all strains into 17 groups, but 15 groups were minor in number (less than 10% of the total strains belonged to each), indicating that the activated sludge bacteria have several common characteristics. For example, most of the isolates were gram negative rods (88%), oxidase-positive (83%), and catalase-positive (91%), while most of them (90%) did not produce acidity from glucose in the Hugh and Leifson medium (Hugh and Leifson, 1953). More interestingly, no fermentative bacteria could be detected in spite of their common presence in the raw wastewater, suggesting their adverse selection in the activated sludge process.

Further physiological tests were carried out, and all the isolates were identified (Table 2.7). As shown in the table, bacteria belonging to genus *Pseudomonas, Flavobacterium,* and the coryneform group (mainly genus *Althrobacter*) dominated, which accounted for more than 10% of the total strains originating from each activated sludge sample. In addition, the presence of small fractions of *Bacillus* sp., *Micrococcus* sp., and unidentified gram positive and negative bacteria were observed.

The dominant bacterial flora of the activated sludge samples tested are expressed as the ratios of bacterial genera or categories as shown in Figure 2.1. From the figure, it can be seen that the bacterial flora of activated sludge cultivated with synthetic wastewater in the laboratory changed considerably with the SRT. Although most bacteria isolated from activated sludges maintained at SRTs of 2 and 5 days were *Pseudomonas* sp., bacteria belonging to *Flavobacterium* and the coryneform group increased with SRT. As a whole, the longer the SRT, the more diverse the bacterial flora. This tendency also seems to be observed for bacterial flora of actual

TABLE 2.6. Characterization of activated sludge bacteria.

Gram Stain	Shape	Motility	Oxidase	Ctalase	Spore	Color	O.F.	Laboratory Unit				Actual Plant	
								SRT: 2-Day	5-Day	10-Day	15-Day	T Plant	S Plant
+	Cocci	–	+	+	–	yellow	–	0	0	0	1	0	0
			–	+	–	red	–	0	0	1	0	0	0
			–	–	–	orange	–	1	0	4	0	0	0
			–	–	–	–	–	0	0	0	0	1	0
	Rod	+	+	+	+	–	–	0	1	1	1	0	0
		–	+	–	–	–	–	0	0	0	0	0	2
	Irregular	–	–	+	–	yellow	–	0	0	2	0	0	0
			–	+	–	–	–	0	6	0	17	2	0
			–	–	–	–	–	4	3	10	1	2	0
–	Rod	+	+	+	–	yellow	O	0	2	0	0	0	0
			+	+	–	yellow	–	3	3	4	0	4	0
			+	+	–	–	O	12	2	5	5	3	2
			+	+	–	–	–	12	41	33	37	14	24
		–	+	+	–	yellow	O	2	0	0	0	1	0
			+	+	–	yellow	–	1	1	8	28	4	2
			+	+	–	–	–	1	1	0	0	0	0
			–	+	–	–	–	0	0	0	0	1	0
							Total	36	60	68	90	32	30

+: Positive, –: negative.
O: acids from glucose under aerobic condition, F: acids from glucose under anaerobic condition.

TABLE 2.7. *Identification of isolated bacterial strains.*

	Laboratory Unit				Actual Plant	
Genus	SRT: 2-Day	5-Day	10-Day	15-Day	T Plant	S Plant
Gram Positive						
Bacillus	0	1	1	1	0	0
Micrococcus	0	0	1	1	0	0
Coryneform group	4	9	12	18	4	0
Unknown (Rod)	0	0	0	0	0	2
(Cocci)	1	0	4	0	1	0
Gram Negative						
Pseudomonas	27	48	42	42	21	26
Flavobacterium	3	1	8	28	5	2
Unknown (Rod)	1	1	0	0	1	0
Total	36	60	68	90	32	30

wastewater treatment plants, judging from the comparison of data between T plant and S plant. Therefore, we can assume that SRT regulation can influence bacterial flora just as the composition of influents, environmental conditions, and other factors do.

As shown in Equation (2.4), SRT control may cause washout of the bacteria with specific growth rates less than the inverse of the SRT. Lower bacterial diversity of activated sludge under a shorter SRT is dependent on the above-mentioned selection in the specific growth rate. This selection seems to have a significant relationship to the trophic levels of bacteria present in activated sludge.

Figure 2.2 shows the changes of populations of bacterial genera or category as SRT changes. Populations were calculated from the data of the viable counts of total heterotrophic bacteria (Table 2.5) and the ratios of each genus or category at certain SRTs (Figure 2.1). This suggests that when the process was held at lower SRTs (2 and 5 days), the TOC-SS loading (the amount of substrates fed into the aeration tank per bacterial cells present), was higher. Therefore, the bacterial populations which can directly utilize the substrates contained in the influent (meat extract and peptone) grew more and dominated the flora. These populations may be called the *primary utilizers,* and *Pseudomonas* sp. are considered the typical genera (McKinney, 1962). On the other hand, the decrease of TOC-SS loading, and selective pressure in specific growth rates at higher SRTs, seemed to cause more competition for substrates between the primary utilizers. This might result in the appearance of secondary utilizers, which can grow on the excreted

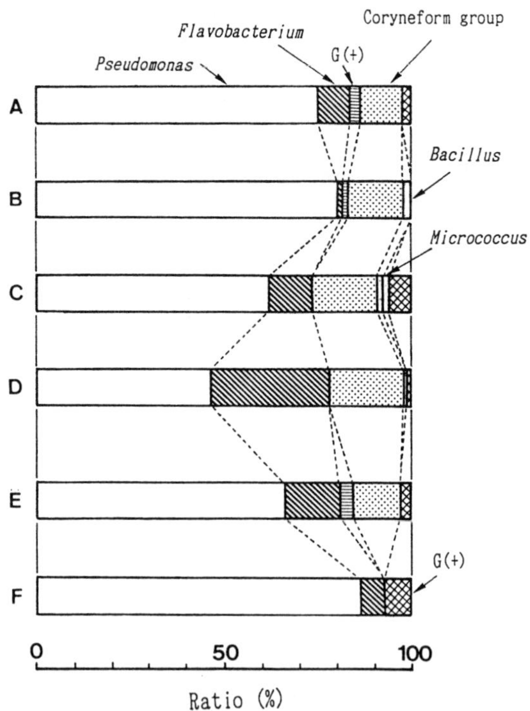

FIGURE 2.1. Bacterial flora of activated sludge samples expressed as ratios of bacterial genera or categories isolated from the activated sludge from laboratory units (A, SRT = 2 days; B, SRT = 5 days; C, SRT = 10 days; D, SRT = 15 days) and actual plants (E, T plant; F, S plant).

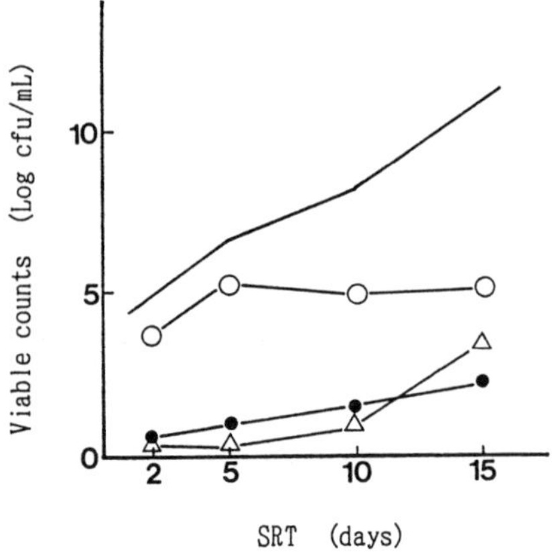

FIGURE 2.2. Changes of bacterial populations by SRT control. Viable counts were calculated from Table 2.5 and Figure 2.1: —, total bacteria; ○, *Pseudomonas;* △, *Flavobacterium;* ●, coryneform group.

TABLE 2.8. Growth and substrate removal properties of representative bacterial strains.

Strain (Identification)	μ(g/g/hr)	ν(g-TOC/g/hr)	TOC Removal (%)
SRT 2-Day			
0206 (*Pseudomonas*)	0.822	0.226	72.7
0217 (*Pseudomonas*)	0.195	0.061	14.9
0232 (Coryneform)	0.125	0.074	7.2
SRT 5-Day			
0508 (Coryneform)	0.121	0.112	11.5
0514 (*Pseudomonas*)	0.166	0.040	23.9
0517 (*Pseudomonas*)	0.439	0.327	55.6
SRT 10-Day			
1011 (*Pseudomonas*)	0.148	n.d.	9.1
1019 (*Pseudomonas*)	0.148	0.063	2.0
1040 (Coryneform)	0.198	0.045	6.9
SRT 15-Day			
1516 (*Flavobacterium*)	0.168	0.197	60.9
1525 (*Pseudomonas*)	0.160	0.128	32.6
1527 (*Pseudomonas*)	0.030	0.035	10.2
1555 (Coryneform)	0.220	0.092	11.5

n.d.: Not determined.

metabolites and the products of cell lysis of the primary utilizers, thus having the advantage of avoiding trophic competition. In this study, bacterial populations belonging to genus *Flavobacterium* and the coryneform group were the putative secondary utilizers which caused the increased diversity of bacterial flora at higher SRTs.

2.3.4 Characteristics of Dominant Bacteria

Three or four representative bacterial strains dominant in the activated sludge flora cultivated with synthetic wastewater at SRTs of 2, 5, 10, and 15 days were selected by the observation of colony and cell morphology and other physiological examinations, and their characteristic growth and substrate removal were investigated. Results are shown in Table 2.8.

In the table, specific growth rate (μ), specific TOC removal rate (ν), and ultimate TOC removal were determined by cultivation with synthetic wastewater (TOC = 400 mg/L) on a rotary shaker (25°C, 120 rpm). μ and ν are defined as follows:

$$\mu = (dX/dt)/X \tag{2.6}$$

$$\nu = (-dTOC/dt)/X \tag{2.7}$$

where X is the cell concentration, TOC is the TOC concentration, and t is time. In this experiment, μ and ν were determined as the mean values observed in the logarithmic growth phase.

As shown in Table 2.8, among the thirteen strains tested, only three strains could remove more than 50% of TOC contained in the synthetic wastewater, while the others could neither grow nor remove TOC rapidly. This suggests that the primary utilizers do not exist in high percentages in activated sludge. TOC removal of the activated sludge process from which the bacterial strains were isolated was more than 90%. However, no representative strains tested could show more than ca. 70% TOC removal. We also investigated TOC removal by mixed cultures composed of 2–4 strains in various combinations. TOC removal was not increased from those observed in the single cultures, suggesting no interactions between the dominant bacteria. Therefore, the roles of minor (not dominant) bacterial populations, and their interactions with dominant bacteria, are considered very important in the complete or efficient removal of organic pollutants in the activated sludge process.

2.4 CONCLUSIONS AND CONSIDERATIONS

In all, 316 bacterial strains isolated from various activated sludge samples were identified as *Pseudomonas* sp., *Flavobacterium* sp., strains of the coryneform group, *Bacillus* sp., *Micrococcus* sp. and unknown gram negative and gram positive rods. These bacterial genera are considered dominant in the activated sludge process. Gram negative bacteria such as *Pseudomonas* and *Flavobacterium* may be the dominant genera in most cases. Therefore, genetic manipulation for aerobic wastewater treatment should involve these bacteria, and further advanced studies must be made on their physiology and genetics. Bacterial strains isolated in such studies may be useful as GEM material for the creation of wastewater treatment bacteria.

SRT manipulation considerably affected the bacterial flora of activated sludge. More such operational parameters need to be determined and investigated, and their influential mechanisms must be made clear before GEM applications can be useful. Knowledge of the bacterial ecology of wastewater treatment processes is important for the assessment of the behaviors and fates of introduced GEMs, their effects on the indigenous bacterial communities, etc. In this study, complicated bacterial interactions in the activated sludge process were suggested—for example, under long SRTs, secondary utilizers were found to dominate primary utilizers in treatment substrates. Such findings may help in the successful development of GEMs.

2.5 REFERENCES

Dias, F. F. and J. V. Bhat. 1964. "Microbial Ecology of Activated Sludge, I. Dominant Bacteria," *Appl. Microbiol.*, 12:412–417.

Harder, W. 1981. "Enrichment and Characterization of Degrading Organisms," in *Microbial Degradation of Xenobiotics and Recalcitrant Compounds,* T. Leisinger, R. Hütter, A. M. Cook and J. Nüesch, eds., London: Academic Press, pp. 77–96.

Hugh, R. and E. Leifson. 1953. "The Taxonomic Significance of Fermentative versus Oxidative Metabolism of Carbohydrates by Various Gram Negative Bacteria," *J. Bacteriol.*, 66:24–26.

McKinney, R. E. 1962. *Microbiology for Sanitary Engineering.* New York: McGraw-Hill Book Co., pp. 139.

Pike, E. B. 1975. "Aerobic Bacteria," in *Ecological Aspects of Used-Water Treatment. Vol. 1,* C. R. Curds and H. A. Hawkes, eds., London: Academic Press, pp. 1–63.

Pike, E. B. and E. G. Carrington. 1972. "Recent Developments in the Study of Bacteria in the Activated-Sludge Process," *Wat. Pollut. Control,* 71:583–600.

Pike, E. B., E. G. Carrington and P. A. Ashburner. 1972. "An Evaluation of Procedures for Enumerating Bacteria in Activated Sludge," *J. Appl. Bacteriol.*, 35:309–321.

Van Gils, H. W. 1964. "Bacteriology of Activated Sludge," report no. 32, Holland, Research Institute for Public Health Engineering.

CHAPTER 3

Bacterial Plasmids in Wastewater Treatment Processes

Little is known about the genetics of wastewater treatment processes. Indeed, it is impossible to wholly elucidate the bacterial genes and their dynamics, since there are many kinds of bacteria with different genetics interacting in complicated ways, as shown in Chapter 2. We must find clues to the genetics of wastewater treatment processes—such as the distribution and behavior of specific and functional genes—before we can produce GEMs and assess their effects, favorable or harmful. Research with bacterial plasmids seems to be a good place to look for clues, since plasmids are much simpler genetic elements than chromosomes, and are therefore more easily studied. Many studies have already been made of the structures and functions of several groups of naturally occurring plasmids, and their important roles in the genetic dynamics of natural environments have been recognized. In wastewater treatment processes, plasmid-mediated gene transfer from GEMs to the indigenous bacteria has become a most significant concern.

This chapter will deal with the bacterial plasmids as an important factor in the application of GEMs in wastewater treatment. The known characteristics of naturally occurring plasmids and their behaviors in natural environments are reviewed first, then the results of research on the distribution of plasmids in a wastewater treatment process (an activated sludge plant) are described as a clue to further studies.

3.1 PLASMIDS

Plasmids are extrachromosomal genetic elements which can replicate independently of the chromosome of the host cell. They are supercoiled double-stranded DNA molecules circular in form, and are present in a wide range of microorganisms, such as bacteria, actinomycetes, yeasts, etc. In general, plasmids detected in bacterial cells are from 1 kilobase (kb = 1,000 nucleotide base pairs of DNA) to over 300 kb in size. Large plasmids greater than 100kb can be found with considerable frequency in gram negative bacteria, whereas gram positive bacteria usually harbor much smaller plasmids. As the size of the bacterial chromosomes is approximately 10^3 kb, a large plasmid may account for a considerable portion of the total genetic information of the host cell, suggesting its significant influences on the existence of the bacterial host. On the other hand, small plasmids seem to have little effect on the general functions of hosts.

In principal, every plasmid possesses the *backbone* or *house-keeping* regions on it (Villarroel et al., 1983). These regions deal with the maintenance, replication, incompatibility, or conjugational transmission of plasmids, and are essential for, or influential in, their existence, continuance, or propagation.

In most cases, plasmids occur in multiple copies in the host bacterial cell, while their copy numbers are determined by the control system of replication. The replication systems of plasmids are classified into two types of control—*stringent control* and *relaxed control.* Stringent control means that the replication of plasmids is associated with the duplication of the bacterial chromosomes of the host, indicating that only one (single copy number) or a few copies (a low copy number) of the plasmid may be maintained in a cell (Novick et al., 1976). Relaxed control means that the plasmids can replicate continuously independently of the duplication of the chromosome, resulting in copy numbers of 10-100 (a medium or high copy number) per cell. In general, small plasmids (<35 kb) may occur at high copy numbers and large plasmids (>35 kb) at single or low copy numbers.

In many cases, different plasmids can be observed coexisting in the same host cell. For example, *Pseudomonas* sp. CF600, which can grow on phenol and a variety of methylated phenols—including cresols and dimethylphenols—harbored two plasmids, one of approximately 45 kb(pVI45) and one of considerably high molecular mass (equal to or more than 150 kb: pVI50), which encoded the complete phenol catabolic pathway (Shingler et al., 1989). The backbone regions of plasmids may determine the combinations of plasmids which can be maintained simultaneously in the same cell. Incompatibility is a phenomenon in which two different plasmids fail to coexist in the same host cell. When one plasmid is introduced into a host which already has other plasmid(s), one plasmid is lost from the cell. Incompatibility is usually observed between plasmids with closely related replication control systems.

Compatibility/incompatibility is a major criterion for classifying bacterial plasmids. In fact, many researchers have classified plasmids into incompatibility groups; any two distinct plasmids belonging to the same incompatibility group cannot coexist in a common bacterial cell. For instance, drug-resistant plasmids detected in *Pseudomonas* can be classified into incompatibility groups P1, P2, P3, P4, N, or W (Chakrabarty, 1976).

Some plasmids occur in a restricted range of host bacteria, while others can be maintained in a wide variety of bacteria, not only in the strains belonging to the same species or genera as the original host, but in other genera as well. These plasmids are termed *broad host range* plasmids. The extent of the host range is determined by the horizontal or vertical transfer (propagation) characteristics of a plasmid. A horizontal transfer plasmid can replicate in different bacterial strains or microorganisms. A vertical transfer plasmid will only replicate in a particular strain of microorganism, and will appear in the daughter cells. The horizontal transfer of genes mediated by plasmids in natural environments will be detailed in this chapter.

In addition to the backbone regions, plasmids contain regions coding additional or extra functions. These functions, and their effects on the fate of host microorganisms in natural environments, will be also described.

3.2 BACTERIAL PLASMIDS IN NATURAL ENVIRONMENTS

3.2.1 Functions and Distribution

Known phenotypic characteristics conferred by naturally occurring plasmids as extra functions, and their examples, can be summarized as follows.

Resistance to antibiotics, heavy metals, etc. Plasmids encoding resistance to agents which have inhibitory or killing effects on the bacteria are termed *R factor plasmids,* and are well-studied and characterized especially in *Escherichia coli* and *Pseudomonas* sp. (Helinski, 1973; Chakrabarty, 1976). R plasmids confer resistance to a wide variety of antibiotics, including penicillin, streptomycin, ampicillin, kanamycin, tetracycline, neomycin, chloramphenicol, erythromycin, sulfonamides, etc. Heavy metals, to which resistance is mediated by plasmids, include mercury, cadmium, lead, bismuth, cobalt, nickel, arsenate, arsenite, etc. As well as *Escherichia coli* and *Pseudomonas* sp., various bacteria have R factor plasmids. For instance, *Staphylococcus aureus* harbored a 3.2-kb plasmid which specified resistance to cadmium (El Solh and Ehrlich, 1982). Multiple resistance to antibiotics and/or heavy metals mediated by a plasmid is common. Typical R factor plasmids RP1 and RK2, which were detected in *Pseudomonas aeruginosa,* encode triple resistance to carbenicillin, kanamycin, and tetracycline (Grinsted et al., 1972; Meyer et al., 1975). *Alcaligenes eutrophus* CH34 isolated from sediments of a decantation basin in a zinc factory harbored two coexisting plasmids, pMOL28 (165 kb) which codes multiple resistance to nickel, chromium, cobalt, and mercury, and pMOL30 (240 kb) which codes resistance to cadmium, cobalt, zinc, mercury, copper, and lead (Diels and Mergeay, 1990). Some typical mercury-resistant plasmids in *Pseudomonas* also show resistance to antibiotics such as kanamycin, streptomycin, sulfonamides, etc. (Chakrabarty, 1976). In addition, resistance to bacteriophages, bacteriocins, and ultraviolet light is also mediated by R factor plasmids.

Degradation of xenobiotic compounds. As a result of numerous studies on the bacterial degradation of xenobiotic compounds, it has been shown that, in many cases, the corresponding catabolic genes are encoded on plasmids. These degradative, or catabolic, plasmids are well-studied in *Pseudomonas,* which have a wide catabolic range of organic compounds. These plasmids are unique in that genes encoding complete or nearly complete degradation of certain compounds (including regulatory genes) are contained all together in one plasmid. For example, toluene/xylene-degrading TOL plasmid pWWO (Williams and Murray, 1974) detected in *Pseudomonas putida* mt-2 contains twelve genes specifying a series of catabolic enzymes, and two regulatory genes. One regulatory gene governs the upper pathway, which transforms toluene/xylene into benzoate/toluate; the second regulatory gene governs the lower pathway, which converts benzoate/toluate into central metabolites. Enormous numbers of xenobiotic compounds can be degraded by the mediation of plasmids, including naphthalene, salicylate, phenol, mandelate, 3-chlorobenzoate, phenoxyacetic acid, 2,4-dichlorophenoxyacetate, p-chlorobiphenyl, p-cymene, nicotine/nicotinate, camphor, n-alkanes (n-octane), 6-aminohexanoate cyclic dimer, etc. Degradation of some sugars, such as lactose, sucrose, raffinose, etc. is also coded on catabolic plasmids. Degradative plasmids sometimes also encode resistance to antibiotics or heavy metals. For example, a species of *Pseudomonas* isolated from soil contained a plasmid encoding both the mineralization of phenylacetate and resistance to mercury (Pickup et al., 1983). *Pseudomonas* isolated from a river also harbored a plasmid specifying degradation of toluene, m-xylene, and p-xylene, as well as resistance to streptomycin and sulfonamides (Yano and Nishi, 1980).

Other functions. Various other functions are known to be coded on naturally occurring plasmids. The production of colicins, a group of antibiotics, was found to be coded on plasmids, such as ColV, detected in *E. coli* strains. Enterobacteria carrying plasmids belonging to this group are of important public health interest, since ColV plasmids increase the pathogenic virulence of certain hosts. Additionally, it has been shown that these plasmids cause the synthesis of excess polysaccharide and clumping, consequently enhancing the attachment of host bacteria to surfaces. This mechanism may protect the bacteria from damage by toxic chemicals (Hicks and Rowbury, 1986). Interactions between bacteria and other organisms are sometimes mediated by plasmids. For example, the crown gall tumor formation in plants by *Agrobacterium* sp., and the nitrogen fixation in legumes by *Rhizobium* sp., are specified by their plasmids (Stotzky and Babich, 1986). Ice nucleic activity in *Pseudomonas viridiflava* KUIN-2, which may cause frost hazard in agricultural crops, was coded on a plasmid pINA, which also specified resistance to ampicillin (Ohgama et al., 1992).

Many plasmids have been detected and isolated from bacterial strains existing in various natural environments, although the functions of a considerable number of these plasmids are not known. These are called *cryptic* plasmids.

The distribution of naturally occurring plasmids, including cryptic ones, and their behaviors (propagation or decay) have not been studied and well elucidated. However, there is some evidence that plasmids are widespread in natural environments, and may play important roles not only in genetics, but also in the ecologies of microbial communities. For example, research which reported plasmid existence in 31% of 155 psychrophilic and psychrotrophic bacteria isolated from sea ice, seawater, sediments, and benthic and ice-associated animals in Antarctica suggested the ubiquitousness of naturally occurring plasmids (Storzky and Babich, 1986). Research on antibiotic-resistant enterobacteria showed an increase of their population associated with the increasing clinical uses of antibiotics (van Elsas, 1992). This research, and the common presence of plasmids in resistant bacteria, suggest significant participation of plasmids in the population changes of microbial communities.

Although the functions and the genetic information coded on plasmids are considered not essential for the host bacteria, they may confer some advantages on their host cells in certain adverse circumstances, such as increasing uses of an antibiotic in a clinic. A degradative plasmid may enable its host to grow predominately in a site polluted with xenobiotic compounds.

3.2.2 Plasmid-Mediated Gene Transfer

The genetic dynamics of bacteria in natural environments are determined by the following factors and/or processes:

- The natural distribution of various genes coded on both chromosomes and plasmids under ordinary circumstances, i.e., the original genetic profile, primarily determines the basic and potential features of the chromosomes and plasmids.
- Selective pressures present in the environment may confer an advantage or disadvantage on certain bacteria; i.e., their genes determine their final ecological niches.
- Gene (DNA) transfer may occur horizontally from primary hosts (donors) to secondary hosts (recipients).

DNA transfer is considered the most important factor and/or process influencing genetic dynamics in natural environments, especially when GEMs are introduced. Three mechanisms of bacterial gene transfer are known—transformation, transduction, and conjugation. In transformation, free DNA molecules are taken directly into bacterial cells. Although this seems very simple, the process can rarely be seen in natural conditions. In transduction, DNA molecules are introduced through infected bacteriophages. When the phage preys on a host, phage DNA is introduced into, and maintained in, the host chromosomes. Conjugation, the only form of bacterial sexual reproduction, requires the contact of bacterial parent and recipient cells. Sex pili produced by the donor move DNA molecules (a whole plasmid and part of a plasmid or chromosome) into the recipient.

Plasmids which can mediate the conjugal transfer are termed *conjugational plasmids* or *self-transmissible plasmids*. They carry genes specifying the production of sex pili, the transfer or mobilization of DNA molecules, surface exclusion properties, etc. As the conjugational plasmids accomodate these genes, their size usually grows to more than 30 kb (large). Many R factor plasmids in *E. coli* and *Pseudomonas* are conjugational and have a broad host range. The RP1 plasmid belonging to incompatibility group P1 in *P. aeruginosa* can be transmitted into gram negative bacteria covering a broad range of species and genera, e.g.: *Pseudomonas fluorescens, E. coli, Salmonella typhimurium, Shigella boydii, Proteus mirabilis, Vibrio cholerae, Acinetobacter calcoaceticus, Azotobacter vinelandii, Neisseria perflava, Rhodospirillum rubrum, Rhodopseudomonas sphaeroides,* etc. (Chakrabarty, 1976). In addition to incompatibility group P1 plasmids, Q and W group conjugational plasmids are known to transfer among various gram negative bacteria (Van Elsas, 1992). On the other hand, several degrading plasmids are conjugational and transmissible among most *Pseudomonas* species, and they cannot be transferred into species of other genera.

In addition, a conjugation-like transfer of plasmids between gram positive bacteria has recently been shown. For example, the plasmid pAMβ1, encoding resistance to erythromycin and lincomycin, moved from *Strepto-*

coccus faecalis to *Bacillus thuringensis* (Lereclus et al., 1983). This type of plasmid may possess the machinery for replication in different hosts in addition to most or all functions required for the normal conjugational transfer. Moreover, recent data have shown the possibility of plasmid transfer from gram positive cocci to a gram negative *E. coli* strain (Brisson-Noel et al., 1988).

In addition to themselves, conjugational plasmids can also transfer nonconjugational plasmids, which cannot transmit themselves. Conjugational and nonconjugational plasmids coexisting in a host bacterial cell can both transfer to a secondary recipient. For example, a nonconjugational plasmid, pKK1, encoding resistance to silver was transferred from a strain of *P. stutzeri* to a strain of *P. putida* after the conjugational plasmid, R68.45 (which was maintained primarily in *P. aeruginosa* PA025), moved into *P. stutzeri* (Haefeli et al., 1981). This type of transfer of nonconjugational plasmids is termed *triparental* conjugation because it requires three members: the donor *P. stutzeri* (pKK1), the recipient *P. putida,* and the helper or mobilizer *P. aeruginosa* PA025(R68.45). Some groups of nonconjugational plasmids can be transferred easily by the mediation of helper plasmids, while other groups are poorly mobilizable. Typical plasmid vectors pBR322 and pBR325 are included in the latter groups.

Additionally, conjugational plasmids sometimes integrate into and mobilize all or part of the host chromosomes. Sex-factor plasmids, which are defined as the plasmids capable of initiating chromosomal gene transfer from one primary cell to another, are known for *Pseudomonas* (Chakrabarty, 1976). Sex-factor plasmids in *P. aeruginosa,* including FP2 and FP39, show a variety of additional phenotypes—such as resistance to mercury—and promote transfer of the chromosome from a single site of origin. R factor plasmids such as R9169 and R6886 can also mobilize the chromosome of *P. aeruginosa* at a high rate. On the other hand, the rate of mobilization of the chromosome of *P. putida* mediated by phage-derived plasmid pfdm, or degrading plasmids such as CAM, TOL, etc., is relatively low. However, typical sex-factor plasmids, such as factor K, initiate transfer of their chromosome at a very high rate.

DNA integration and mobilization mechanisms mediated by conjugational plasmids can be also observed in interactions among plasmids. Kivisaar et al. showed the construction of phenol-degrading plasmids from multiple plasmids through the mobilization of plasmid-coded genes (Kivisaar et al., 1990). *Pseudomonas* sp. EST1001, a derivative of *Pseudomonas* sp. S13 isolated from soil, harbored three plasmids and utilized phenol as a sole carbon source. Although these three plasmids were not coded individually for the complete degradation of phenol, the selection of the naturally occurring derivative of this strain under the presence of phenol resulted in the appearance of mutants harboring a single plasmid. The resultant plasmids, pEST1005 and pEST1226, were both able to degrade phenol. This suggests the integration of DNA fragments controlling phenol degradation from the original three plasmids into the single plasmids. Thus, the rearrangement of degrading or resistant genes coded on the plasmids is considered common, and plasmid-associated genetic evolution has been recognized (Shahrabadi et al., 1975; Williams and Worsey, 1976).

Moreover, transposable elements, such as transposons and insertion sequences (IS), are often carried in plasmids, and may play an important role in the evolution of bacterial genes (Young, 1992). Transposable elements are the main nucleotide sequences in a replicon (chromosome or plasmid) which can be moved to another replicon without losing a functionality.

Although the mechanisms of bacterial gene transfer mediated by plasmids have been well-researched, there is a dearth of observations in natural environments because of the difficulties in studying complicated mixed microbial communities in situ.

3.2.3 Plasmid Transfer in Wastewater Treatment Processes

Plasmid transfer, especially the transfer of recombinant plasmids, in wastewater treatment processes, as well as in other natural and seminatural environments, such as lake water (Fulthorpe and Wyndham, 1991), soils (Ramos-Gonzalez et al., 1991), drinking water (Sandt and Herson, 1991), etc., has been demonstrated by a few research groups.

In their research, Gealt et al., (1985), concluded that the transfer of man-made DNA sequences coded on the recombinant plasmids will occur only if the GEMs are released into the wastewater treatment processes.

First, they confirmed the transfer of nonconjugational plasmid vectors pBR322 and pBR325 from the laboratory strains of *E. coli* (donor) to bacterial strains indigenous to raw wastewater by the mediation of mobilizer plasmid R100-1 maintained in laboratory *E. coli* strains. Plasmid transfer was observed during a 25-hour coincubation of these strains both in L broth and sterilized wastewater, suggesting the presence of indigenous recipient bacteria (Gealt et al., 1985).

Next, the mobilizer strains capable of transferring pBR325 to both laboratory and natural bacterial strains of *E. coli* were isolated from raw wastewater. Several indigenous strains of enteric bacteria could transfer not only their antibiotic-resistant plasmids (or phenotypes), but also pBR325 maintained in laboratory *E. coli* strains to plasmid-free recipients (McPherson and Gealt, 1986).

Finally, they demonstrated the mobilization of plasmid pHSV from *E. coli* HB101 to both laboratory and indigenous *E. coli* strains in the model wastewater treatment facility. Plasmid transfer occurred under several different treatment conditions with the presence of mobilizer strains. However, wastewater treatment microorganisms were excluded in this experiment (Mancini et al., 1987).

Mach and Grimes (1982) also described the transfer of R factor plasmid in the filter diffusion chamber loaded in the primary and final sedimentation tank. This in situ study was, however, also carried out without the presence of the indigenous bacteria, which were removed with a membrane filter. Resistant enteric bacteria were isolated and mated with recipient bacteria (*E. coli* and *Shigella sonnei* strains) in the chamber, resulting in the occurrence of R plasmids.

Plasmid transfer from the GEM under the presence of activated sludge microcosms was investigated by McClure et al., (1991). *P. putida* UWC1 containing 3-chlorobenzoate degrading nonconjugational recombinant plasmid pD10 and an activated sludge were mated in the filter, and the 28 transconjugants which acquired pD10-like plasmids were isolated. Interestingly, these transconjugants were all gram negative. The results suggested the presence of natural mobilizers and recipients indigenous to activated sludge, as was demonstrated for raw wastewater by Gealt et al. (1985).

More recently, real evidence was demonstrated for plasmid transfer from a GEM into a secondary recipient after the GEM was introduced into an activated sludge process; the GEM contained a conjugational plasmid. The donor, *P. putida* KT2440(pWWO-EB62), carrying a conjugational plasmid, and the recipient, *P. putida* UWC1, were introduced into a model activated sludge unit, and their survival and the appearance of transconjugants were investigated. After 2 days, the formation of transconjugants was observed. However, when the nonconjugational plasmid pFRC20P was maintained in the donor, *Pseudomonas* sp. B13FR1, and a similar experiment was carried out, there were no transconjugants detected (Nusslein et al., 1992).

This research emphasizes the importance of naturally occurring plasmids in wastewater treatment processes.

3.3 BACTERIAL PLASMIDS IN A WASTEWATER TREATMENT PROCESS

3.3.1 Plant Description

Research on the distribution of naturally occurring plasmids in S wastewater treatment plant (S plant) was carried out. Figure 3.1 shows a

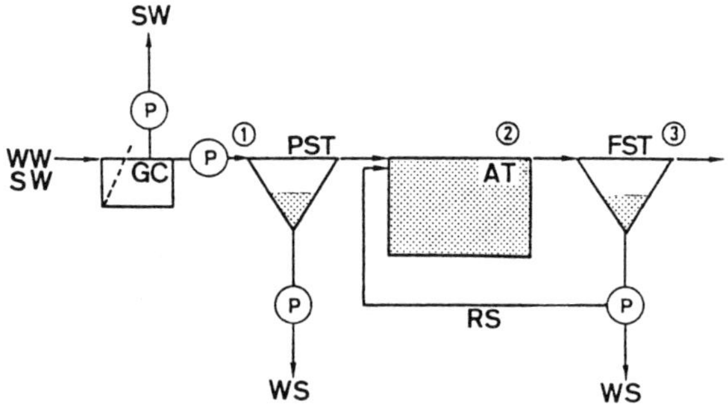

FIGURE 3.1. Schematic diagram of wastewater flow in S plant. WW, wastewater; SW, stormwater; WS, waste sludge; RS, return sludge; P, pump; GC, grit chamber; PST, primary sedimentation tank; AT, aeration tank; FST, final sedimentation tank. Samples were collected at ① (influent), ② (activated sludge), and ③ (effluent).

schematic diagram of the wastewater flow in S plant. This plant is an activated sludge plant (step aeration) with a maximum capacity of 52,560 m^3/day. The combined sewerage collects wastewater and stormwater from an area covering approximately 10 km^2. At heavy rains, flows of wastewater and rainwater are discharged from the grit chamber without treatment. Most wastewater treated in this plant came from domestic use (population of approximately 95,000), but there was also some commercial wastewater. The hydraulic retention times of the primary sedimentation tank, aeration tank, and final sedimentation tank were approximately 2–3 hours, 4–5 hours, and 2–3 hours, respectively, and the SRT of the process was maintained at approximately 5 days. Effluent from the final sedimentation tank is discharged to K river after 20–30 min chlorination.

Samples for isolation of bacterial strains, which were examined for plasmid contents, were taken from the overflow of the grit chamber (influent sample), mixed liquor from the aeration tank (activated sludge sample), and overflow of the final sedimentation tank (effluent sample). The points at which these samples were collected are indicated in Figure 3.1. The concentrations of SS of the influent, activated sludge, and effluent samples used were 62.3, 944, and 9.2 mg/L, respectively. As soon as possible after they were collected, the samples were plated onto solid media to obtain bacterial strains. The activated sludge was tested after sonication, which was necessary to release bacterial cells from flocs, while the influent and effluent were tested after homogenization by a vortex mixer.

3.3.2 Plasmid Screening Method

A well-established way to screen for the presence of plasmids is to extract plasmid DNA from bacterial cells, and by use gel electrophoresis to prepare a visual analysis. Although there are a number of plasmid extraction procedures which vary in subtle ways, they all have three main stages – lysis of bacterial cells, selective release of plasmid DNA from other cell matrices (cell debris, proteins, lipids, chromosomal DNA, ribonucleic acid [RNA], etc.), and removal of such contaminants to recover plasmid DNA. Wastewater and activated sludge samples contain many kinds of bacteria, harboring various plasmids. Screening methods must be rapid and simple, and must detect many kinds of bacteria and plasmids. In our research, we used five different procedures, slightly modified, to screen for plasmid-harboring bacterial strains.

Kado and Liu (KL) method. A cell pellet was suspended in E-buffer (40 mM Tris, 2 mM EDTA; ph = 7.9) and lysed by adding SDS and NaOH. The lysate was heated at 50–60°C for 20 min to facilitate elimination of the chromosomal DNA and RNA, and then emulsified with unbuffered phenol-chloroform (1:1). The phenol-chloroform mixture denatures and extracts proteins and cell debris. It also lowers the pH of the lysate, so the chromosomal DNA denatured by alkaline SDS treatment renatures and aggregates to form an insoluble network. The emulsion was broken by centrifugation, and an upper aqueous phase was obtained (phenol-chloroform extraction). Plasmid DNA contained in the aqueous phase was precipitated with 1/10 volume of 3M sodium acetate and 2 volumes of ethanol. This operation is termed *ethanol precipitation* (Kado and Liu, 1981).

Birnboim and Doly (BD) method. Cells were treated with lysozyme to weaken the cell wall, and then lysed completely with SDS and NaOH in a buffer containing Tris, EDTA and glucose. Afterward, 1/2 volume of high salt acid solution (3M sodium acetate; pH = 4.8) was added to the lysate. In this step, the lysate was neutralized, and the chromosomal DNA aggregated. Simultaneously, the high salt concentration caused precipitation of protein SDS complexes and high molecular weight RNA. Precipitates (proteins, chromosomal DNA, RNA, etc.) were removed by centrifugation. The precipitates were also subjected to phenol-chloroform extraction and RNAse treatment to completely remove proteins and RNA. Plasmid DNA was recovered by ethanol precipitation (Birnboim and Doly, 1979).

Casse et al. method. Cells were lysed with SDS and NaOH in TE buffer (50 mM Tris, 20 mM EDTA) by gentle mixing for 20–25 min at 34°C. The pH of the lysate was lowered to 8.5–8.9 with 2M Tris buffer and kept at ambient temperature for 2 hours with gentle mixing. NaCl was added to a final concentration of 3%. After 30 min., the lysate was extracted with phenol, then plasmid DNA was recovered by ethanol precipitation. This method was developed originally for isolating large plasmids from *Rhizobium meliloti* (Casse et al., 1979).

Boiling method. For lysis, cells were suspended in sucrose-Triton solution (8% sucrose, 0.5% Triton X-100, 50 mM EDTA, 10 mM Tris; pH =

8.0) and lysozyme was added to a final concentration of 1 μg/mL. The solution was placed in a boiling water bath for 1 min and cell debris (pellet) was discarded with a toothpick or by centrifugation. Boiling releases plasmid DNA from the chromosome or lysed cells. Plasmid DNA in the supernatant was precipitated with isopropanol (Kadowaki, 1980).

TELT method. Cells were suspended in TELT solution (4% Trotin X-100, 62.5 mM EDTA, 2.5 M LiCl, 50 mM Tris; pH = 8.0) for lysis, then phenol-chloroform was added and the mixture was agitated. The solution was centrifuged and the upper aqueous phase was subjected to ethanol precipitation. This method is simple to use for rapid screening of plasmids (Kadowaki, 1980).

Bacterial strains used in this comparative study were natural *E. coli* and *P. putida* harboring plasmids pUC19 (Vieira and Messing, 1982), pBH500 (Fujita et al., 1991), and RP4 (Datta et al., 1971) which are approximately 2.7 kb, 17.5 kb and 56.4 kb in size, respectively. Crude plasmid preparation samples were subjected to electrophoresis and analyzed by the Meyers et al. method, except that the 0.7% agarose gels were run at 100 V for 20–30 min on a horizontal gel electrophoresis apparatus, Mupid-3, which is a product of Advance Co., Tokyo, Japan (Meyers et al., 1976). Gels stained with ethidium bromide were observed under ultraviolet light to detect bands of plasmid DNA.

Photographs of gel electrophoresis of crude plasmids extracted by these five methods are compared in Figure 3.2, and the results are summarized in Table 3.1. Plasmids were detected from all strains as clear bands when extracted by the KL method and the BD method. Successful and reproducible plasmid extractions from all strains could not be accomplished with the other three methods. In particular, pBH500 in *P. putida* BH could be detected only by the BD and KL methods. Crude preparation by the BD method often contained much chromosomal DNA and RNA, which prevented the clear detection of plasmid bands as backgrounds in gel electrophoresis, whereas the RNAse treatment reduced the effect of contaminated RNA as shown in Figure 3.2. However, the BD method tended to extract plasmids in both cc form and oc form, which produced two plasmid bands in the electrophoresis of the extract from a bacterium harboring only one plasmid. Both the BD and KL methods were considered suitable for this study.

The BD and KL methods were used to screen plasmids from several wastewater bacterial strains. The strains tested were obtained from wastewater samples collected from the main sewer at Osaka University (Suita Campus) in May, 1992 and November, 1992. In all, 30 antibiotic-resistant bacteria were examined for the presence of plasmids, using both methods. Although the KL method detected 16 plasmids with sizes between 1.8–168 kb from 7 strains, no clear plasmid bands were observed with the

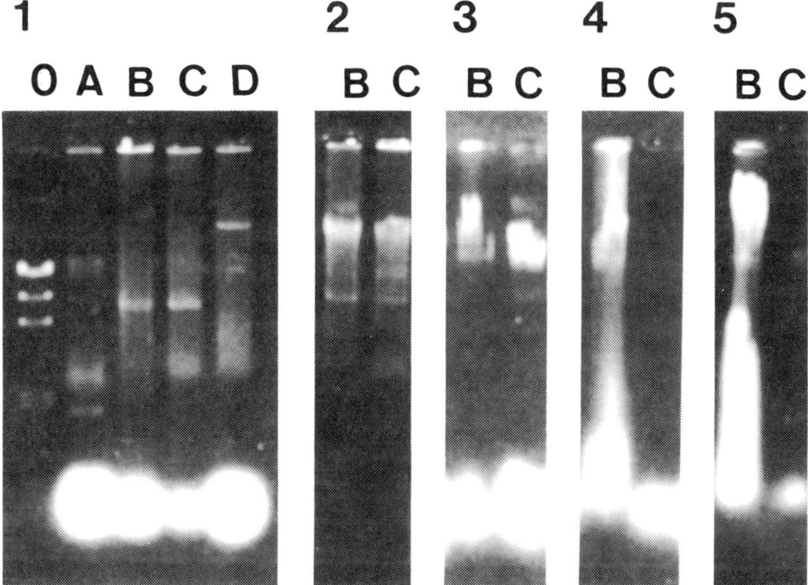

FIGURE 3.2. Gel electropheresis of crude plasmid extract prepared by various methods. Plasmids were extracted by the method of Kabo and Liu (1), Birnboim and Doly (2), Casse et al. (3) the boiling method (4), and the TELT method (5) from JM103(pUC19) (A), BH(pBH500) (B), C600(pBH500) (C), and C600(RP4) (D). Lane O, *Hind* III-digested λ DNA. Reprinted from *Water Research, Vol. 27,* Fujita, Ike and Suzuki, "Screening of Plasmids from Wastewater Bacteria," p. 950, 1993. With kind permission from Pergamon Press, Ltd., Headington Hill Hall, Oxford OX3 0BW, UK.

BD method because of the strong background, which may have been due to contaminated chromosomal DNA, and protein or cell debris.

Therefore, the KL method was adopted as the most suitable plasmid screening method. However, when the original KL method was used, plasmid bands observed in gel electrophoresis tended to be faint, and procedures had to be carried out on a large scale (maximum volume = 9 mL). Therefore, the method was modified slightly by condensing plasmid DNA as an ethanol precipitate to allow operations on a smaller scale. This modified protocol is shown in Figure 3.3.

3.3.3 Distribution of Plasmids

The distribution of plasmids in the S wastewater treatment plant was estimated by screening representative bacterial strains isolated from the influent, activated sludge, and effluent using the modified KL method shown

TABLE 3.1. Comparison of plasmid extraction methods.

Method	Plasmid band[a]				Time[b]	Backgrounds[c]
	A	B	C	D		
Kado and Liu (KL)	○	○	○	○	1.5 hr	none or faint
Birnboim and Doly (BD)	○	○	○	○	2.5 hr	RNA, chromosomal DNA
Casse et al.	○	–	○	○	1 day	chromosomal DNA
Boiling method	○	–	–	–	1.5 hr	protein, debris
TELT method	○	–	–	–	1.0 hr	protein, debris

[a]○: Plasmid band was observed, –: not observed from A: JM103 (pUC19), B: BH (pBH500), C: C600 (pBH500), and D: C600 (RP4).
[b]Time necessary for preparation of crude plasmid extract.
[c]Backgrounds which disturbed the detection of plasmid bands in the electrophoresis.
Reprinted from *Water Research, Vol. 27,* Fujita, Ike and Suzuki, "Screening of Plasmids from Wastewater Bacteria," p. 951, 1993, with kind permission from Pergamon Press, Ltd., Headington Hill Hall, Oxford, OX3 0BW, UK.

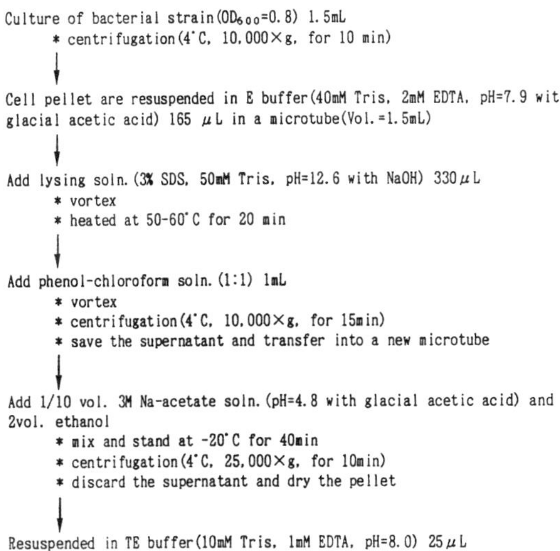

FIGURE 3.3. Plasmid preparation procedure by a modified Kado and Liu method. Reprinted from *Water Research, Vol. 27,* Fujita, Ike and Suzuki, "Screening of Plasmids from Wastewater Bacteria," p. 950, 1993. With kind permission from Pergamon Press Ltd., Headington Hill Hall, Oxford OX3 0BW, UK.

in Figure 3.3. Bacterial strains examined for plasmids were of three groups – dominant heterotrophic bacteria, enteric bacteria, and antibiotic-resistant bacteria. The medium used to isolate dominant bacteria was CGY agar (see Section 2.3.2). Desoxycholate agar (Eiken Chemical Co., Tokyo, Japan) was used to isolate enteric bacteria. Antibiotic-resistant bacteria were obtained by using CGY agar supplemented with antibiotics, which were added at these concentrations: ampicillin, 100 mg/L; Kanamycin, 80 mg/L; tetracycline, 25 mg/L; and streptomycin, 25 mg/L.

Viable plate counts of antibiotic-resistant bacteria in the influent, activated sludge, and effluent samples are listed in Table 3.2. Counts of total heterotrophic bacteria (on CGY agar) and enteric bacteria (on desoxycholate agar), respectively, are also shown in the table. In the influent and effluent, a small percentage of the total bacteria was counted in the media supplemented with a single antibiotic agent, excepting tetracycline. Counts were 2.4–4.3 $\times$ 10^5 cfu/mL for the influent, and 3.0–6.0 $\times$ 10^3 cfu/mL for the effluent. However, the counts with tetracycline were an order of magnitude smaller than those observed with the other antibiotics. Bacteria expressing triple resistance to ampicillin, kanamycin, and tetracycline were 0.02% and 0.06% of the total in the influent and effluent. On the other hand,

*TABLE 3.2. Viable counts of antibiotic-resistant bacteria in S plant.**

Media	Sample		
	Influent	Activated sludge	Effluent
CGY agar	5.3×10^{6}	1.0×10^{8}	1.3×10^{5}
Desoxycholate agar	3.2×10^{5}(6.04)	9.8×10^{4}(0.10)	3.0×10^{3}(2.31)
CGY agar + AP	2.4×10^{5}(4.53)	1.3×10^{5}(0.13)	3.8×10^{3}(2.92)
CGY agar + Km	4.3×10^{5}(8.11)	2.2×10^{5}(0.22)	6.0×10^{3}(4.62)
CGY agar + Tc	6.4×10^{3}(0.12)	1.2×10^{4}(0.01)	1.8×10^{2}(0.14)
CGY agar + Sm	2.5×10^{5}(4.72)	1.4×10^{6}(1.40)	5.2×10^{3}(4.00)
CGY agar + Ap, Km, Tc	7.8×10^{2}(0.015)	3.2×10^{3}(0.003)	7.7×10^{1}(0.06)

*Viable counts are expressed as cfu/mL.
Ap: ampicillin, Km: kanamycin, Tc: tetracycline, Sm: streptomycin.

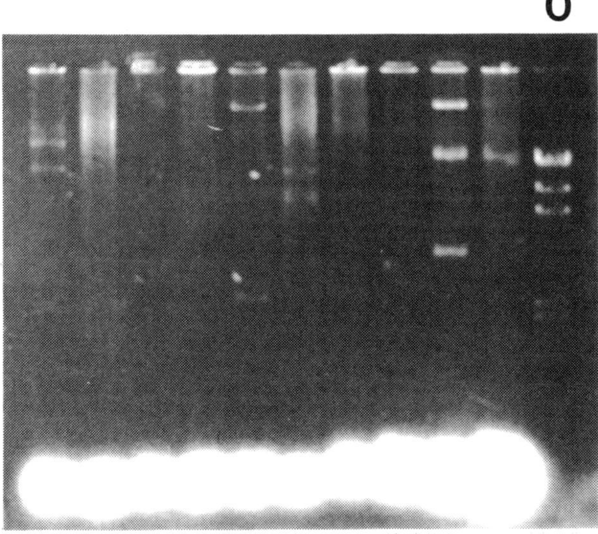

FIGURE 3.4. An example of plasmid screening. Lane 0. *Hind* III-digested λ DNA.

the ratios of resistant bacteria in the activated sludge were much lower than those in the influent and effluent.

Each of the 10 morphologically different colonies which appeared on the media was isolated (70 strains total for each sample), and its plasmid contents investigated. An example of gel electrophoresis for plasmid screening is shown in Figure 3.4. We identified the clear band as a plasmid. Table 3.3 summarizes the results of this plasmid screening. All bacterial strains harboring at least one plasmid detected in this study are listed in Table 3.4.

Among the 70 bacterial strains isolated from the influent, activated sludge, and effluent, 78, 21, and 51 plasmids were detected in 23 (32%), 9 (13%), and 20 (29%), respectively. In all, 150 plasmids were detected in 52 (25%) of 210 strains tested. We tried to isolate plasmid-harboring bacteria by screening out strains with antibiotic resistance as a selective marker. (Resistant genes are common on plasmids). Plasmid-harboring bacteria were isolated at slightly higher rates on CGY agar (33%) and desoxycholate agar (43%)—neither of which contained any antibiotic—than on antibiotic-supplemented media (17% and 27%, respectively).

Even though all possible antibiotics were not used in this study, this result suggests that plasmids which code for functions other than antibiotic resistance might exist in wastewater and activated sludge.

Sixteen of 52 plasmid-harboring strains had only one plasmid. The rest harbored multiple (2–9 distinct) plasmids (numbers of strains with 2, 3, 4,

TABLE 3.3. *Plasmid screening in S plant.*

Sample	Media†	No. of bacteria with plasmids* (plasmid bands)	No. of bacteria without plasmids
Influent	CGY agar	4(24)	6
	Desoxycholate agar	4 (11)	6
	CGY agar + Ap	2 (6)	8
	CGY agar + Km	3 (8)	7
	CGY agar + Tc	3 (13)	7
	CGY agar + Sm	3 (8)	7
	CGY agar + Ap, Km, Tc	4 (8)	6
	total	23 (78)	47
Activated sludge	CGY agar	1 (2)	9
	Desoxycholate agar	3 (10)	7
	CGY agar + Ap	1 (1)	9
	CGY agar + Km	3 (7)	7
	CGY agar + Tc	0 (0)	10
	CGY agar + Sm	1 (1)	9
	CGY agar + Ap, Km, Tc	0 (0)	10
	total	9 (21)	61

TABLE 3.3. (continued).

Sample	Media†	No. of bacteria with plasmids* (plasmid bands)	No. of bacteria without plasmids
Effluent	CGY agar	4 (11)	6
	Desoxycholate agar	6 (22)	4
	CGY agar + Ap	2 (5)	8
	CGY agar + Km	2 (2)	8
	CGY agar + Tc	2 (4)	8
	CGY agar + Sm	2 (2)	8
	CGY agar + Ap, Km, Tc	2 (5)	8
	total	20 (51)	50
	Total	52 (150)	158

*Results of plasmid screening from each of 10 strains.
†Media used for isolation of bacterial strains.

TABLE 3.4. Bacterial strains harboring plasmids.

Origin	Media	Strain	Plasmid No.	Plasmid size (kb)
Influent	CGY	ICGY–2	5	89.9, 13.8, 6.87, 4.75, 2.45
		ICGY–3	6	25.0, 8.35, 4.64, 3.68, 3.20, 2.45
		ICGY–4	4	123, 87.0, 6.40, 3.03
		ICGY–9	9	103, 67.6, 48.1, 12.4, 8.60, 4.04, 3.47, 2.80, 2.32
	Desoxycholate	IDES–1	3	44.4, 5.62, 3.55
		IDES–3	5	46.5, 8.12, 5.15, 3.49, 2.74
		IDES–6	1	3.65
		IDES–7	2	4.43, 3.49
	CGY + Ap	IA–4	2	91.2, 8.53
		IA–7	4	4.75, 4.59, 4.21, 3.02
	CGY + Km	IK–2	4	129, 39.3, 3.53, 2.48
		IK–4	3	129, 42.9, 3.53
		IK–5	1	147
	CGY + Tc	IT–1	3	9.51, 5.38, 3.37
		IT–2	8	22.1, 12.3, 7.51, 7.17, 6.47, 6.20, 4.59, 3.80
		IT–3	2	2.91, 1.79
	CGY + Sm	IS–1	6	31.9, 19.6, 4.36, 3.37, 2.82, 2.24
		IS–3	1	7.44
		IS–7	1	131
	CGY + Ap, Km, Tc	IAKT–1	2	123, 29.5
		IAKT–5	2	44.0, 22.5
		IAKT–7	3	123, 35.9, 2.89
		IAKT–9	1	4.62

TABLE 3.4. (continued).

Origin	Media	Strain	Plasmid No.	Plasmid size (kb)
Activated sludge	CGY	SCGY-5	2	114, 38.1
	Desoxycholate	SDES-3	3	56.5, 3.03, 2.74
		SDES-5	5	53.8, 7.20, 4.21, 3.06, 2.29
		SDES-9	2	7.20, 4.65
	CGY + Ap	SA-8	1	16.0
	CGY + Km	SK-1	1	106
		SK-2	3	103, 31.2, 3.15
		SK-8	3	124, 25.7, 16.7
	CGY + Sm	SS-2	1	16.0
Effluent	CGY	ECGY-2	2	7.24, 4.13
		ECGY-3	5	31.4, 19.2, 6.77, 4.04, 3.12
		ECGY-7	3	104, 33.8, 7.24
		ECGY-10	1	6.86
	Desoxycholate	EDES-1	5	53.8, 32.8, 6.56, 3.89, 3.49
		EDES-2	5	92.9, 44.4, 14.0, 10.8, 4.56
		EDES-3	2	47.6, 30.7
		EDES-4	2	3.43, 2.80
		EDES-7	6	40.5, 15.2, 6.07, 4.89, 4.01, 3.34
		EDES-9	2	34.2, 2.97
	CGY + Ap	EA-4	4	50.7, 5.11, 2.68, 2.28
		EA-8	1	105
	CGY + Km	EK-4	1	10.0
		EK-5	1	117
	CGY + Tc	ET-1	3	8.77, 5.11, 2.85
		ET-3	1	5.11
	CGY + Sm	ES-2	1	2.37
		ES-8	1	8.91
	CGY + Ap, Km, Tc	EAKT-2	4	140, 61.7, 35.9, 80.8
		EAKT-9	1	4.47

5, 6, 8, and 9 plasmids were 11, 9, 5, 6, 3, 1, and 1, respectively). Since plasmids coexisting in a host cell can be classified into different incompatibility groups, the common presence of multiple plasmids in bacterial strains suggests the existence of various types of plasmids in wastewater and activated sludge. The coexistence in indigenous bacteria of plasmids from different incompatibility groups is thought to have some effect on gene transfer or mobilization. For example, plasmids belonging to the same incompatibility group of a conjugational plasmid may prevent its transfer into their own host.

Distribution of bacterial plasmids in an activated sludge wastewater treatment plant could be estimated from the results of plasmid screening of dominant bacteria isolated on CGY agar.

In our study, 4 of 10 dominant bacteria in the influent possessed multiple plasmids, demonstrating the widespread presence of plasmids in bacterial populations of domestic wastewater as well as in other aquatic environments (see Section 3.2.1). On the other hand, among the dominant bacteria isolated from the activated sludge, only two plasmids were detected, both in the same strain. This suggests that in the aeration tank, bacteria without plasmids are capable of growing faster or decaying more slowly than plasmid-harboring bacteria. This further suggests that plasmids give little or no advantage to the host. The ratio of plasmid-harboring bacteria isolated from the effluent was higher (4 of 10 strains) than the ratio in the activated sludge, suggesting the increase or selective survival of plasmid-harboring bacteria in the final sedimentation tank. It also seemed that the increase in the ratio of plasmid-harboring bacteria in the effluent depended on plasmid transfer mediated by conjugational plasmids.

The tendency for plasmid-harboring bacteria to decrease in the aeration tank and increase in the final sedimentation tank was observed not only for dominant bacteria, but also for other antibiotic-resistant populations. The exception to this rule was organisms resistant to kanamycin. Figure 3.5 shows the distribution of plasmids of various bacterial populations in S plant.

3.3.4 Sizes of Detected Plasmids

The 150 plasmids detected in the screening tests ranged in size from 1.8–147 kb. They were classified into five size catagories: smaller than 5 kb, 5–10 kb, 10–30 kb, 30–100 kb, and larger than 100 kb (see Figure 3.6). Small plasmids, below 5 kb, can contain only a few genes, as the average gene is approximately 0.9 kb (Stotzky and Babich, 1986). Large plasmids, greater than 30 kb, are large enough to accommodate genes to code the pro-

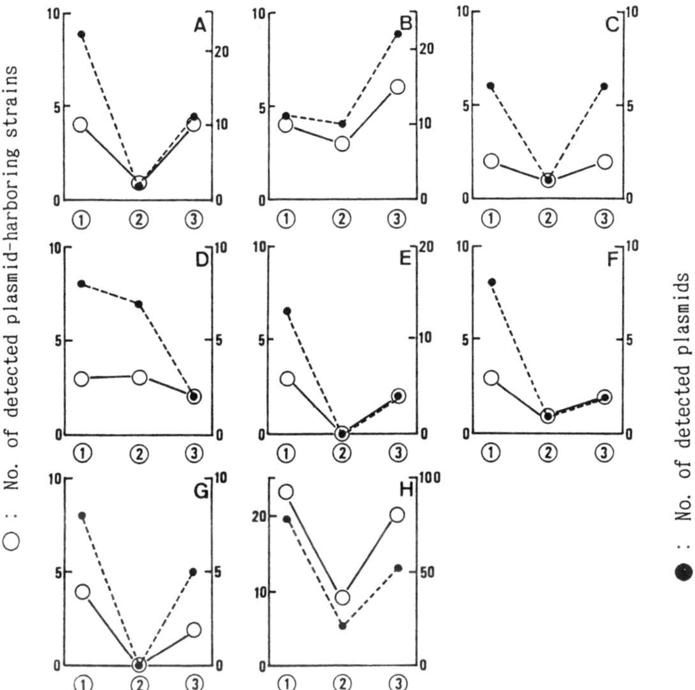

FIGURE 3.5. Plasmid distribution in S plant. Plasmids were detected in each 10 of bacterial strains isolated from influent①, activated sludge②, and effluent③ by using CGY agar (A), desoxycholate agar (B), and CGY agar supplemented with ampicillin (C), kanamycin (D), tetracycline (E), streptomycin (F), and the last three antibiotics combined (G). Distribution of all plasmids is summarized as shown in H.

duction of sex pili and the mobilization of DNA molecules, and they can be conjugational.

Dominant bacteria harbored mainly small (<5 kb) plasmids (38%). Large plasmids (>30 kb), including huge plasmids (>100 kb), were also detected at a high percentage (30%). This suggests that a certain portion of dominant bacteria in wastewater and activated sludge have conjugational plasmids. In enteric bacteria, and ampicillin- and streptomycin-resistant bacteria, small plasmids (<5 kb) were dominant (46–50%) as they were in dominant bacteria. By contrast, large plasmids (>30 kb) accounted for 61% and 54% in the bacterial populations which had resistance to kanamycin and triple resistance to ampicillin, kanamycin, and tetracycline, respectively. However, large plasmids were not found in the tetracycline-resistant bacteria, which contained plasmids mainly from 5–10 kb (54%).

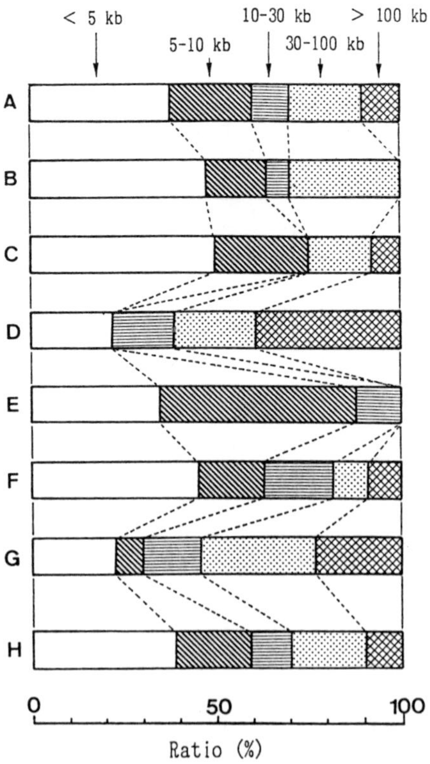

FIGURE 3.6. Size distribution of plasmids. Plasmids were detected in each 30 of bacterial strains isolated by using CGY agar (A), desoxycholate agar (B), and CGY agar supplemented with ampicillin (C), kanamycin (D), tetracycline (E), streptomycin (F), and the last three antibiotics (G). Size distribution of all plasmids is summarized as shown in H.

3.4 CONCLUSIONS AND CONSIDERATIONS

Plasmid screenings from wastewater and activated sludge bacteria indicate that plasmid-harboring bacteria are widespread in wastewater treatment processes. Plasmids were between 1.8–147 kb in size, and seemed to fall into at least nine different incompatibility groups. The coexistence of nine distinct plasmids in a bacterial strain, suggests the existence of a wide variety of plasmids in their backbones. Large plasmids, which could be conjugational in size, were found in several dominant bacteria. It may be concluded that the plasmids indigenous to wastewater treatment processes have considerable effect on the behavior of certain bacterial populations and genes, including GEMs which may be introduced accidentally or deliberately.

Therefore, further studies on the functions of plasmids, more precise quantitative analysis of plasmid distribution, etc. are necessary for predicting the fate of GEMs and their recombinant DNA in wastewater treatment processes.

3.5 REFERENCES

Birnboim, H. C. and J. Doly. 1979. "A Rapid Alkaline Extraction Procedure for Screening Recombinant Plasmid DNA," *Nucleic Acids Res.*, 7:1513–1523.

Brisson-Noel, A., M. Arthur and P. Courvalin. 1988. "Evidence for Natural Gene Transfer from Gram-Positive Cocci to *Escherichia coli*," *J. Bacteriol.*, 170:443–448.

Casse, F., C. Boucher, J. S. Julliot, M. Michel and J. Denarie. 1979. "Identification and Characterization of Large Plasmids in *Rhizobium meliloti* Using Agarose Gel Electrophoresis," *J. Gen. Microbiol.*, 113:229–242.

Chakrabarty, A. M. 1976. "Plasmids in *Pseudomonas*," *Ann. Rev. Genet.*, 10:7–30.

Datta, N., R. W. Hedge, E. J. Shaw, R. B. Sykes and M. H. Richmond. 1971. "Properties of an R Factor from *Pseudomonas aeruginosa*," *J. Bacteriol.*, 108:1244–1249.

Diels, L. and M. Mergeay. 1990. "DNA Probe-Mediated Detection of Resistant Bacteria from Soils Highly Polluted by Heavy Metals," *Appl. Environ. Microbiol.*, 56:1485–1491.

El Solh, N. and S. D. Erlich. 1982. "A Small Cadmium Resistant Plasmid Isolated from *Staphylococcus aureus*," *Plasmid*, 7:77–84.

Fulthorpe, R. R. and R. C. Wyndham. 1991. "Transfer and Expression of the Catabolic Plasmid pBRC60 in Wild Bacterial Recipient in a Freshwater Ecosystem," *Appl. Environ. Microbiol.*, 57:1546–1553.

Fujita, M., M. Ike and S. Hashimoto. 1991. "Feasibility of Wastewater Treatment Using Genetically Engineered Microorganisms," *Wat. Res.*, 25:979–984.

Gealt, M. G., M. D. Chai, K. B. Alpert and J. C. Boyer. 1985. "Transfer of Plasmid pBR322 and pBR325 in Wastewater from Laboratory Strains of *Escherichia coli* to Bacteria Indigenous to the Waste Disposal System," *Appl. Environ. Microbiol.*, 49:836–841.

Grinsted, J., J. R. Saunders, L. C. Ingram, R. B. Sykes and M. H. Richmond. 1972. "Properties of an R Factor Which Originated in *Pseudomonas aeruginosa* 1822," *J. Bacteriol.*, 110:529–537.

Haefeli, C., C. Franklin and K. Hardy. 1984. "Plasmid-Determined Silver Resistance in *Pseudomonas stutzeri* Isolated from a Silver Mine," *J. Bacteriol.*, 158:389–392.

Helinski, D. R. 1973. "Plasmids Determined Resistance to Antibiotics: Molecular Properties of R Factors," *Ann. Rev. Microbiol.*, 27:437–470.

Hicks, S. J. and R. J. Rowbury. 1986. "Virulence Plasmid-Associated Adhesion of *Escherichia coli* and Its Significance for Chlorine Resistance," *J. Appl. Bacteriol.*, 61:209–218.

Kado, C. I. and S. T. Liu. 1981. "Rapid Procedure for Detection and Isolation of Large and Small Plasmids," *J. Bacteriol.*, 145:1365–1373.

Kadowaki, Y. 1980. "Preparations of Plasmid DNA," in *Saiboukougaku-Jikkenn Protocol.* Research Institute of Medical Science ed., Tokyo: Tokyo Univ., pp. 71–86.

Kivisaar, M. A., R. Horak, L. Kasak, A. L. Heinaru and J. K. Habicht. 1990. "Selection of Independent Plasmid Determining Phenol Degradation in *Pseudomonas putida* and the Cloning and Expression of Genes Encoding Phenol Monooxygenase and Catechol 1,2-dioxygenase," *Plasmid,* 24:25–36.

Lereclus, D., G. Menou and M. M-. Lecadet. 1983. "Isolation of a DNA Sequence Related to Several Plasmids from *Bacillus thuringensis* after a Mating Involving the *Streptococcus faecalis,*" *Mol. Gen. Genet.*, 191:307–313.

Mach, P. A. and D. J. Grimes. 1982. "R-Plasmid Transfer in a Wastewater Treatment Plant," *Appl. Environ. Microbiol.*, 44:1395–1403.

Mancini, P., S. Fertels, D. Nave and M. A. Gealt. 1987. "Mobilization of Plasmid pHSV106 from *Escherichia coli* HB101 in a Laboratory-Scale Waste Treatment Facility," *Appl. Environ. Microbiol.*, 53:665–671.

McClure, N. C., A. J. Weightman and J. C. Fry. 1991. "Survival of *Pseudomonas putida* UWC1 Containing Cloned Catabolic Genes in a Model Activated-Sludge Unit," *Appl. Environ. Microbiol.*, 55:2627–2634.

McPherson, P. and M. A. Gealt. 1986. "Isolation of Indigenous Wastewater Bacterial Strains Capable of Mobilizing Plasmid pBR325," *Appl. Environ. Microbiol.*, 51:904–909.

Meyer, R., D. Figurski and D. R. Helinski. 1975. "Molecular Vehicle Properties of the Broad Host Range Plasmid RK2," *Science,* 190:1226–1228.

Meyers, J. A., D. Sanchez, L. P. Elwell and S. Falkow. 1976. "Simple Agarose Gel Electrophoretic Method for the Indentification and Characterization of Plasmid Deoxyribonucleic Acid," *J. Bacteriol.*, 127:1529–1537.

Novick, R. P., R. C. Clowes, S. N. Cohen, R. Curtiss III, N. Datta and S. Falkow. 1976. "Uniform Nomenclature for Bacterial Plasmids: A Proposal," *Bacteriol. Rev.*, 40:168–189.

Nusslein, K., D. Maris, K. Timmis and D. F. Dwyer. 1992. "Expression and Transfer of Engineered Catabolic Pathways Harbored by *Pseudomonas* spp. Introduced into Activated Sludge Microcosms," *Appl. Environ. Microbiol.*, 58:3380–3386.

Ohgama, H., Y. Hasegawa, H. Obata and T. Tokuyama. 1992. "Ice-Nucleation Gene of *Pseudomonas viridiflava* KUIN-2 Coded on a Plasmid," *Biosci. Biotech. Biochem.*, 56:1690–1691.

Pickup, R. W., R. J. Lewis and P. A. Williams. 1983. "*Pseudomonas* sp. MT14, a Siol Isolate Which Contains Two Large Plasmids, One a TOL Plasmid and One Coding for Phenylacetate Catabolism and Mercury Resistance," *J. Gen. Microbiol.*, 129:153–158.

Ramos-Gonzalez, M. -I., E. Duque and J. L. Ramos. 1991. "Conjugational Transfer of Recombinant DNA in Cultures and in Soils: Host Range of *Pseudomonas putida* TOL Plasmids," *Appl. Environ. Microbiol.*, 57:3020–3027.

Sandt, C. H. and D. S. Herson. 1991. "Mobilization of the Genetically Engineered Plasmid pHSV106 from *Escherichia coli* HB101 (pHSV106) to *Enterobacter cloacae,*" *Appl. Environ. Microbiol.*, 57:194–200.

Shahrabadi, M. S., Bryan, L. E. and H. M. Van den Elzen. 1975. "Further Properties of P-2-R Factors of *Pseudomonas aeruginosa* and Their Relationship to Other Plasmid Groups," *Can. J. Microbiol.*, 21:592–605.

Shingler, V., F. C. H. Franklin, M. Tsuda, D. Holroyd and M. Bagdasarian. 1989. "Molecular Analysis of a Plasmid-Encoded Phenol-Hydroxylase from *Pseudomonas* CF600," *J. Gen. Microbiol.*, 135:1083–1092.

Stotzky, G. and H. Babich. 1986. "Survival and Genetic Transfer by Genetically Engineered Bacteria in Natural Environments," *Advances in Appl. Microbiol.*, 31:93–138.

Van Elsas, J. D. 1992. "Antibiotic Resistance Gene Transfer in the Environment: An Overview," in *Genetic Interactions among Microorganisms in the Natural Environment*, E. M. H. Wellington and J. D. Van Elsas, eds., New York: Pergamon Press, pp. 17–39.

Vieira, J. and J. Messing. 1982. "The pUC-Plasmid, An MBmp7-Derived System for Insertion Mutagenesis and Sequencing with Synthetic Universal Primers," *Gene*, 19:295–268.

Villarroel, R., R. W. Hedges, R. Maenhaut, J. Leemans, M. Engler, M. Van Montague and J. Schell. 1983. "Heteroduplex Analysis of P-Plasmid Evolution: The Role of Insertion and Deletion of Transposable Elements," *Mol. Gen. Genet.*, 189:390–399.

Williams, P. A. and K. Murray. 1974. "Metabolism of Benzoate and Methylbenzoates by *Pseudomonas putida (arvilla)* mt-2: Evidence for the Existence of a TOL Plasmid," *J. Bacteriol.*, 120:416–423.

Williams, P. A. and M. J. Worsey. 1976. "Ubiquity of Plasmids in Coding for Toluene and Xylene Metabolism in Soil Bacteria," *J. Bacteriol.*, 125:818–828.

Yano, K. and T. Nishi. 1980. "pKJI, a Naturally Occurring Conjugative Plasmid Coding for Toluene Degradation and Resistance to Streptomycin and Sulfonamides," *J. Bacteriol.*, 143:552–560.

Young, J. P. W. 1992. "The Role of Gene Transfer in Bacterial Evolution," in *Genetic Interactions among Microorganisms in the Natural Environment*, E. M. H. Wellington and J. D. Van Elsas, eds., New York: Pergamon Press, pp. 3–13.

CHAPTER 4

Genetic Manipulation Techniques

Sophisticated tools and techniques have been developed to break through the barriers between the theory and practice of genetic manipulation. To breed GEMs applicable to wastewater treatment, desirable genes from natural environments must be cloned, and the cloned genes introduced into host microorganisms. Present technology does not allow the *in vitro* synthesis of desirable genes and microorganisms. For example, say we want to remove xenobiotic compounds from wastewater with a low pH. We would start with bacteria that can degrade the target compounds at low pH, clone the gene responsible for the process, and introduce that gene to the bacterial host.

This chapter will deal with genetic manipulation techniques, especially methods of gene cloning and DNA introduction to hosts. Cloning systems which seem useful for breeding wastewater treatment bacteria will be discussed. Also, as a practical example, the cloning of phenol degrading genes from *P. putida* BH and their analyses are detailed. Since most engineers and researchers concerned with wastewater treatment know something about genetic engineering, only basic and general procedures are included here. If further information and specialized protocols for gene cloning are necessary, many introductory books and manuals are available (Chakrabarty, ed., 1987; Glover, ed., 1985; Maniatis et al., 1982).

4.1 GENERAL PROCEDURES

4.1.1 Gene Cloning

Gene cloning means to isolate a desired functional gene as a DNA fragment, which is why it is also called DNA cloning. The average genome size of a bacterium is 10^3 kb. For example, the chromosomes of *P. aeruginosa*, *P. putida*, and *P. fluorescence* are 3624, 3820, and 4107 kb in size (Holloway and Morgan, 1986). The average number of nucleotide base pairs contained in a gene is 900 (Stotzky and Babich, 1986). The technique involves the selection of a specific gene (DNA fragment) from the many genes coded on the genome (chromosome and plasmids, if present) of the donor cell.

Generally, gene cloning follows these steps:

- The DNA fragment which contains the target gene is dissected from the total DNA of the donor cell.
- The dissected DNA fragment is joined to a cloning vehicle (vector) which can replicate or propagate in a recipient.
- The cloning vehicle, including the desired DNA fragment, is transferred into a recipient cell.
- Cells which harbor the cloning vehicle, with or without the desired DNA fragment, are isolated or selected.
- From these cells, the desired recombinants carrying the target DNA fragment are screened.

The cloned gene (DNA fragment) can be isolated from complex mixtures of DNA molecules and propagated. The desired gene and its products can be recovered in large quantity and high purity. Moreover, cloned genes can be introduced into a variety of recipient organisms and can express their functions in those organisms. Genotypes which do not exist in nature can be created, and the gene function *in vivo* may be broadened. GEM techniques (which cannot be performed by conventional physical or chemical methods) can be applied not only to bacterial genes, but also to genes originating in any other organism.

4.1.2 DNA Manipulation

The primary technique in gene cloning is the recombination of DNA molecules, in which hybrid DNA molecules from multiple parents (donors) are created through cutting and joining.

DNA molecules are usually cut with site-specific restriction endonuclease. Although appropriate shear force can be used to produce randomly sized dissected fragments, DNA fragments cleaved at defined points are more useful for gene cloning. Site-specific restriction endonucleases are enzymes which catalyze the cleavage of DNA within or near to specific nucleotide sequences. Their typical recognition sites are 4 to 6 nucleotides in length, and show a twofold axis of symmetry. Some endonuclease recognition sequences and their points of cleavage are shown in Figure 4.1. As shown in the figure, many enzymes make staggered cuts which can be easily

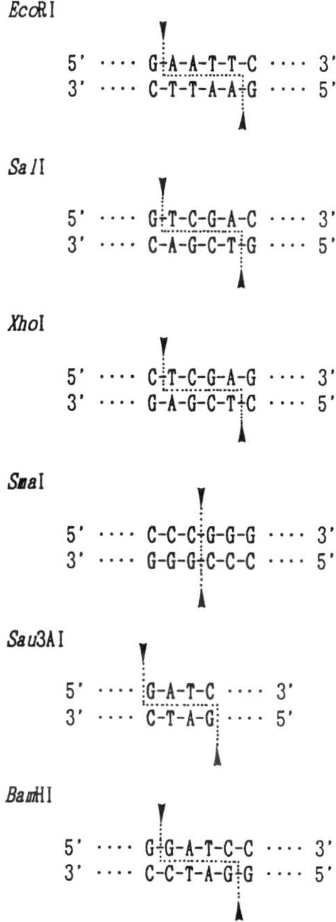

FIGURE 4.1. Recognition sites for restriction endonucleases. Dotted lines indicate the points of cleavage.

rejoined with the protruding cohesive 5′ termini or 3′ termini (e.g., *Eco*RI, *Sal* I), while others make blunt ends (e.g., *Sma* I) with cleavage at the axis of symmetry.

A DNA fragment for gene cloning should be large enough to carry the complete gene, but not so large that it can't be joined to a cloning vehicle. Therefore, restriction endonucleases which have recognition sites composed of six base pairs are generally used. If bases constituting DNA molecules were distributed at random, digestion with endonucleases would make fragments with an average size of 4.1 kb. In actual DNA digestions, the average size varies—*Bam*HI and *Sal* I produced the DNA fragments with an average size of 6 and 8 kb, respectively, while others generated smaller fragments (Murray, 1978).

DNA digestion with endonucleases is performed in buffers with ionic strengths that differ according to the enzymes (Table 4.1). The DNA and enzyme are contained in 20 μL or less of buffer, and kept at an optimal temperature for the required time. The reaction can be stopped by adding EDTA to a final concentration of 10 mM.

All digested DNA fragments can be recovered in purified form by one phenol/chloroform extraction, one chloroform extraction, and ethanol precipitation (see Section 3.3.2). Sizing, separation, identification, and purification of the DNA fragments also can be carried out by agarose gel electrophoresis. As shown in Figure 3.2 and Figure 3.4 (in Section 3.3.2 and Section 3.3.3, respectively), DNA molecules of different sizes can be detected directly in gel at different locations, because the distance a DNA molecule migrates is determined by its molecular weight. Many methods for recovering DNA fragments from agarose gels have been developed, such as electrolysis into dialysis bags or troughs, onto a dialysis membrane, etc. When low-melting point agarose gels are used, DNA can be recovered by phenol/chloroform extraction, and ethanol precipitation of dissolved gels (Maniatis et al., 1982).

DNA molecules with corresponding cohesive 5′ and 3′ termini may be easily joined through hydrogen bonds. However, the unions must be secured by covalent bond. The DNA ligase, which catalyzes the formation of a phosphodiester bond between adjacent 3′-OH and 5′-P termini in DNA, can be used for this purpose. This is diagrammed in Figure 4.2. Since DNA ligase commonly exists in bacterial hosts, ligation may occur *in vivo*. However, *in vitro* ligation before DNA is introduced to hosts is more efficient.

In practice, the litigation reaction catalyzed by T4 ligase is performed in a ligation buffer containing Tris, $MgCl_2$, dithiothreitol, and ATP, since the enzyme requires ATP and Mg^{2+} as cofactors. Another type of ligase, obtained from *E. coli,* requires NAD as a cofactor. The reaction is carried out in 20 μL or less volume of mixture of DNA molecules to be joined (each at 500 μg/mL). First, the mixture is heated to 56–60°C for 5 min to produce

TABLE 4.1. Buffers for restriction endonuclease treatment.

Buffer (Ion Strength)	NaCl	Tris	$MgCl_2$	Dithiothreitol
Low buffer	0 mM	10 mM	10 mM	1 mM
Medium buffer	50 mM	10 mM	10 mM	1 mM
High buffer	100 mM	50 mM	10 mM	1 mM

Ligation of Chohesive End

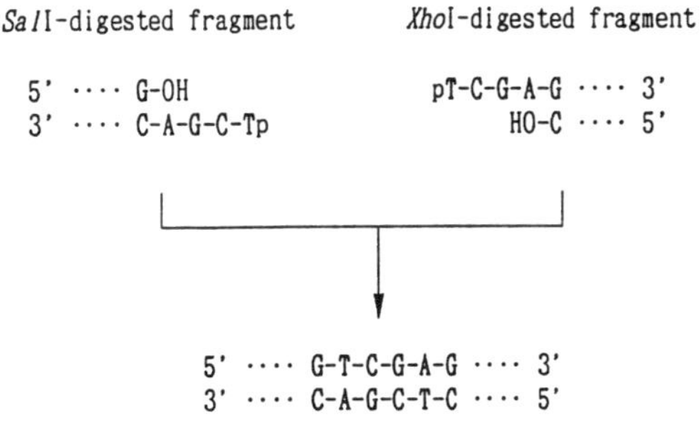

Ligation of Blunt End

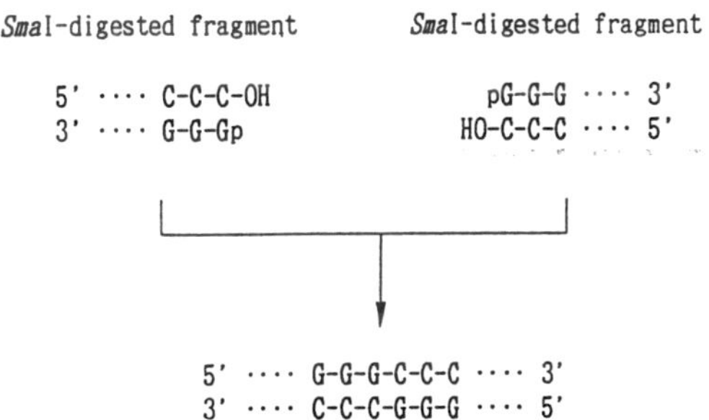

FIGURE 4.2. Joining DNA fragments with DNA ligase.

single-stranded molecules, followed by cooling in ice. Then, 1 unit of ligase per microgram of DNA is added and kept at 16°C for a long enough time for a sufficient reaction to occur. This reaction is considerably affected by the concentration of different forms of DNA and their termini.

Normally, the ligation is performed between DNA molecules which have complementary cohesive ends. Many site-specific restriction endonucleases produce specific cohesive 5′ or 3′ termini. Therefore, DNA fragments digested by a certain endonuclease can be easily joined through their unique and identical nucleotide sequences, and covalently sealed by DNA ligase. Interestingly, there are several different enzymes which produce the same or complementary cohesive ends. For example, DNA fragments digested by *Sal*I (G↓TCGAC) can be ligated to those generated by *Xho*I (C↓TCGAG), as shown in Figure 4.2. More interestingly, the resultant hybrid sites cannot be cleaved by either *Sal* I nor *Xho*I.

Another procedure for producing complementary ends is to add the defined single-strand nucleotide sequences, a defined sequence is attached to one DNA fragment and its complementary sequence to another. The terminal transferase, which catalyzes the addition of deoxynucleotides to the 3′-OH end of DNA molecules, can be successfully used for this purpose. This enzyme can produce single-stranded homopolymeric tails on double-stranded DNA in the presence of a single type of nucleotide triphosphate. Poly dA tails, which are added to one DNA fragment by terminal transferase with ATP, and poly dT tails, which are attached to another fragment in the same manner (except with TTP), can be easily joined through their complementary tails. More than 40 nucleotide sequences (homopolymeric tails) make the joint quite stable. When Mg^{2+} is added as a cofactor, the terminal transferase extends the single-stranded DNA from a protruding 3′-terminus; when CO^{2+} is also present, DNA can be added to blunt ends of *Sma*I-digested DNA fragments or 5′-termini.

A variety of enzymes, such as site-specific restriction endonuclease, T4 DNA ligase, terminal deoxynucleotide transferase, etc. are used in DNA manipulation. These enzymes are supplied by the manufacturer in concentrated form and/or as kits, with detailed explanation for protocols in practical applications. Other enzymes often used include alkaline phosphatase—which catalyzes the removal of 5′-terminal phosphates from DNA and is used in the inactivation of cohesive ends—and DNA polymerase, which produces double-stranded DNA from single-stranded DNA, and is used mainly to label DNA.

4.1.3 Introduction of DNA Molecules into Bacterial Hosts

Among the three gene transfer processes mentioned in Section 3.2.2, transformation and transduction are mainly used to introduce recombinant DNA molecules into bacterial host strains. In general, cloned genes are joined to cloning vehicles, then transferred into host bacteria. Cloning vehicles should be able to:

- acquire foreign genes through simple manipulations, and to efficiently introduce the genes into a well-characterized bacterial host
- replicate autonomously in a host, even after the foreign DNA fragments are joined to them
- be easily separated from other background DNA which may originate from a host

Cloning vehicles which have been developed for *E. coli* strains, which are the most studied and developed hosts, are bacterophages (λ, M13-derivatives, etc.) and plasmids including cosmids.

Most sophisticated phage vectors are derived from λ, which is a double-stranded DNA virus with a genome size of approximately 50 kb. Figure 4.3 shows a schematic diagram of the life cycle of λ. Soon after linear λ DNA in the phage particle is injected into a bacterial cell, it circulates, then shows alternately lytic and lysogenic growth. During lytic growth, the DNA is replicated and transcribed by host enzymes, consequently many phage particles are synthesized, ultimately causing cell lysis. During lysogenic growth, phage genome is integrated into the chromosome and replicated as a part of the host genome. Consequently, if a foreign DNA fragment can be inserted into λ DNA, it can be introduced into *E. coli* hosts and remain stable during lysogenic growth. During lytic growth, the cloned gene propagates and can be recovered as phage particles. A phage vector λgt10 is also shown in Figure 4.3. As this vector contains a single *Eco*RI site, a *Eco*RI-digested foreign DNA fragment can be inserted here. The resultant recombinant phage DNA is packaged into the prehead, then the complete functional phage particle is constructed *in vitro* by adding a series of

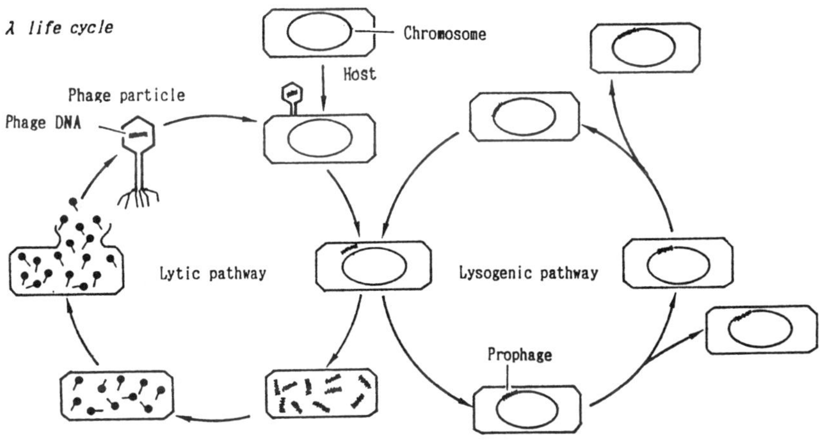

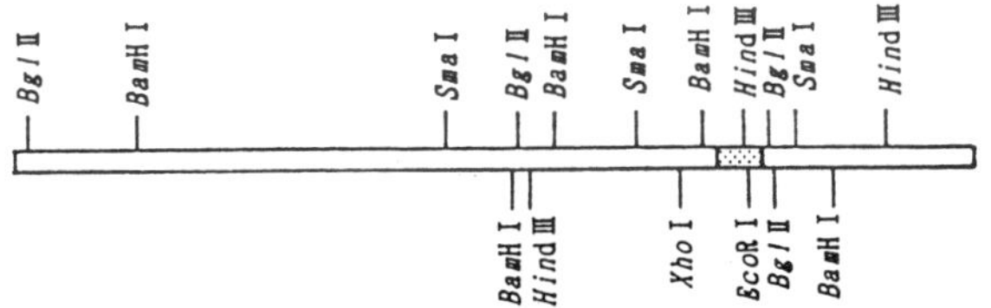

FIGURE 4.3. Life cycle of bacteriophage λ and a phage vector λgt10.

corresponding enzymes and proteins. A host strain is infected with this recombinant phage, and the foreign genes are introduced into it by transduction.

Plasmid vectors are well-developed and versatile cloning vehicles. The biology of plasmids was described in Chapter 3, and their derived vectors are detailed in the next section, so only a general procedure for gene cloning with plasmid vectors is described here. Plasmid vectors, which are double-stranded circular DNA molecules, can be cleaved in at least one restriction endonuclease recognition site lying in the region not needed for replication and maintenance, and the foreign DNA fragment can be inserted there. The recombinant plasmids can be introduced into a host strain by transformation. For *E. coli*, various plasmid transformation techniques have been developed that have a high rate of success and are simple to operate (Hanahan, 1985). *Pseudomonas* strains can be also transformed with plasmids with similar techniques (Chakrabarty et al., 1975; Bagdasarian et al., 1981).

Conventional and general procedures for bacterial transformation depend on the bacterial ability to take up free DNA molecules when the cells are treated with $CaCl_2$ and/or other agents ($MgCl_2$, RbCl, etc.). Such treated cells are called *competent* cells. In practice, bacterial cells in their exponential growth phase are washed once, then resuspended in an ice-cold buffer containing $CaCl_2$ (e.g., 10 mM Tris [ph = 8], 50 mM $CaCl_2$ for *E. coli*), and maintained at 4°C for 12–24 hours to obtain competent cells. Transformation is performed by adding the plasmid DNA to the competent cells and maintaining the mixture at 0°C for 30 min and then heating quickly to 42°C for 2 min (heat shock). Transformed cells recover their activity during a 30 min to 1 hr incubation in a rich medium. They are then plated onto selective media to screening for recombinants. The important factors influencing the transformation rate are the growth phase of cells and the temperature and time of the $CaCl_2$ treatment. If necessary, competent cells can be stored at −70°C. Recombinant plasmids introduced into a host replicate autonomously, and can be recovered by plasmid extraction methods. These steps for gene cloning in plasmid vectors are diagrammed in Figure 4.4.

Cosmid vectors are unique plasmid vectors in that they can be packaged into λ phage and thus introduced into bacterial hosts by transduction, which has a higher rate of gene transfer success than transformation. Moreover, since they can survive and replicate as plasmids in host strains, it is easy to propagate and separate cloned DNA fragments. A schematic diagram of gene cloning with a cosmid vector is shown in Figure 4.5. An example of the use of the cosmid vector in gene cloning is described in Section 4.3.

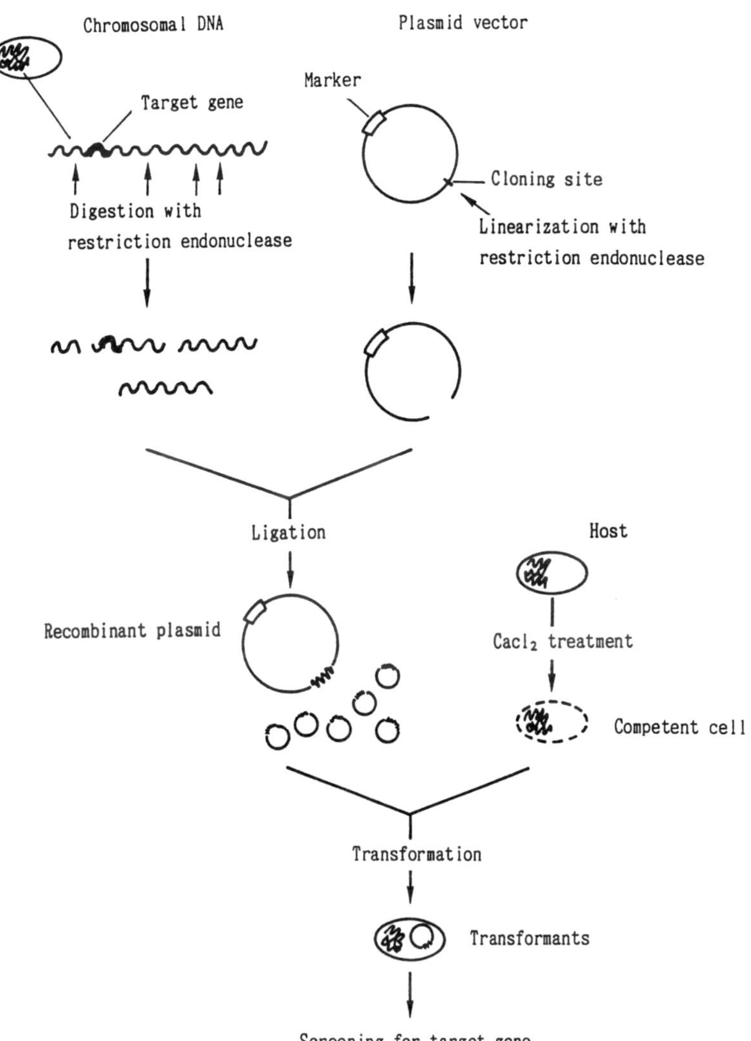

FIGURE 4.4. Gene cloning in a plasmid vector.

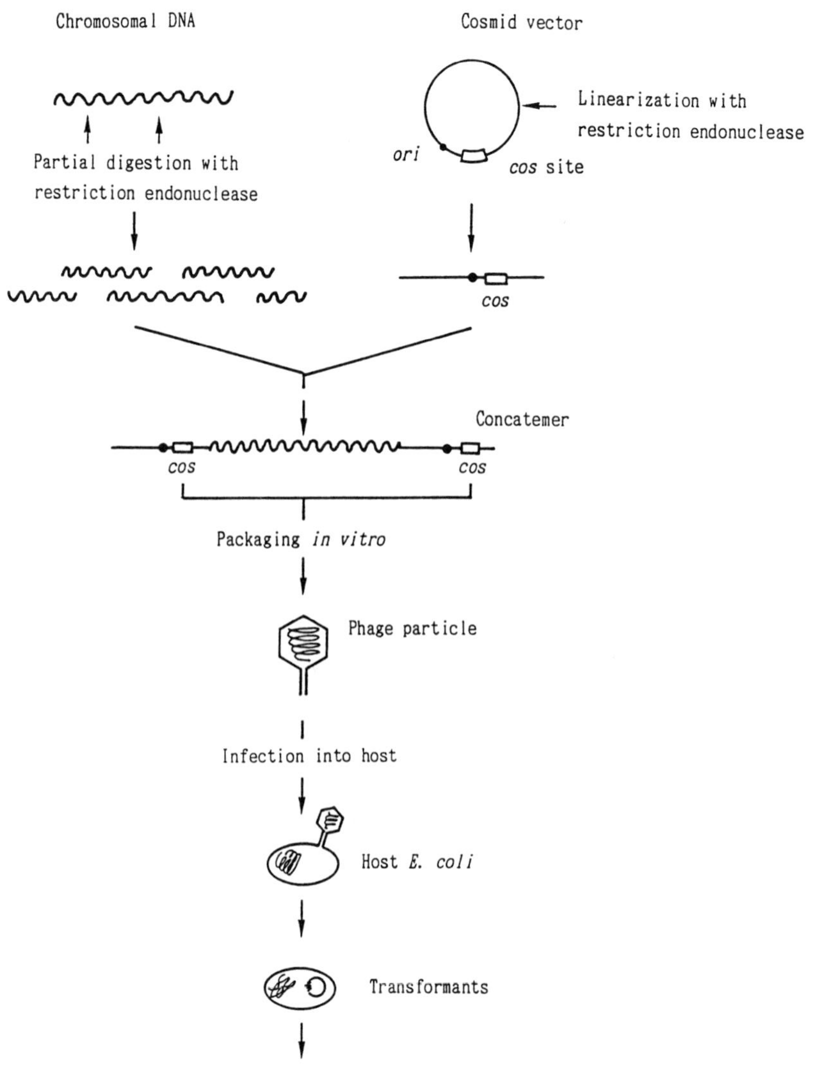

FIGURE 4.5. Gene cloning in a cosmid vector.

4.1.4 Screening Recombinants

Clones obtained in the normal process of gene cloning have different DNA fragments from the source DNA and from each other. Cells carrying the target gene(s) must be screened and isolated from a randomly constructed "gene library." The probability $P(\%)$ that a gene library includes a desired DNA fragment is expressed as follows (Clark and Carbon, 1976):

$$P = 100 \times \{1 - ((1 - A/L)F)^n\} \qquad (4.1)$$

where A is size of desired DNA segment or gene, L is average size of cloned DNA fragments (DNA fragments inserted into cloning vehicles), F is fraction of genome represented by an average fragment, and n is number of clones constructed.

For example, Clark and Carbon calculated that 720 transformants were necessary to achieve a 90% certainty that one *E. coli* gene was contained in a gene library constructed by using a plasmid vector carrying DNA fragments with average molecular weight of 8.5×10^6. Doubling the size of this gene library raises the certainty to 99% (Clark and Carbon, 1976). At any rate, a considerable number of transformed clones must be screened to obtain the desired genes.

Since cloning vehicles are generally coded with more than one distinguishable marker—such as antibiotic-resistance—clones harboring cloning vehicles with or without inserted foreign DNA fragments can easily be isolated on appropriate media. Plasmid vector pBR322 (Bolivar et al., 1977), for example, codes for resistance to ampicillin and tetracycline, and hosts can be isolated directly on media containing these antibiotics. In practical gene cloning, the ligation of cloning vehicles and foreign DNA fragments may produce "self-ligated" cloning vehicles in addition to recombinant DNA fragments. Self ligation means that the cleaved sites of the vehicle DNA are rejoined without the insertion of foreign DNA fragments. Therefore, before screening for target genes, we need to know how much DNA was inserted into the cloning vehicles. This can be done by investigating selective markers which are activated or inactivated by DNA insertion at the sites present in the region coding for markers. For example, when DNA fragments are cloned into the *Sal*I site of pBR322, transformants carrying the recombinant plasmids become sensitive to tetracycline, so they can be isolated phenotypically.

In principal, final screening of the specific target clones cannot be performed by any of the previously described methods, so a suitably simple and reliable procedure should be chosen from among the various possible strategies. The desired clone or gene can then be detected phenotypically or genotypically.

Detection by phenotypes is very simple. We have described an example of such detection of the transformants carrying the catechol 2,3-oxygenase (C23O) encoding gene cloned from *P. putida* BH (Fujita et al., 1991). Since C23O catalyzes the conversion of catechol into 2-hydroxymuconic semialdehyde (2-HMS), the clone harboring the desired gene was identified by spraying the clone colonies with catechol solution—positive colonies turned yellow due to the formation of 2-HMS. However, if the cloned gene cannot be expressed sufficiently in a host cell (expression of the desired gene may vary considerably with the cloning vehicle and host bacterial strain), or has no phenotypically detectable markers, such methods cannot be applied.

In that case, detection by genotypes, such as hybridization techniques of DNA or RNA, is available. These methods use the labeled DNA or RNA fragment which is complementary to the target gene sequence (probe) as a marker. In a colony hybridization procedure, transformed colonies are transferred onto nitrocellulose filters and treated with alkali to lyse cells and denature the double-stranded DNA to single-stranded; denatured DNA remains bound to the filter. The washed filters are hybridized with an appropriate DNA or RNA probe, and the hybridized colonies can be detected by radioautography or fluorography. When probes can be obtained, these methods are reliable for detection of the target gene.

4.2 CLONING SYSTEMS FOR WASTEWATER TREATMENT

4.2.1 Development of Plasmid Vectors

For breeding wastewater treatment bacteria, cloning systems applicable to a wide variety of genes and hosts seem necessary. Plasmid vectors which have been developed for various purposes are considered suitable for such applications. Essential or desirable properties for plasmid vectors are as follows. (Those which are most essential were mentioned in Section 4.1.3):

- They should be relatively small, so that unnecessary regions of plasmids are excluded, and so that they are easily handled with little damage. This may also lead to efficient transformation.
- They should replicate readily (see Section 3.1) so that a large amount of plasmid DNA with inserted foreign DNA can be recovered.
- They should have a single recognition site for a variety of restric-

tion endonucleases, into which foreign DNA fragments digested with corresponding restriction enzymes can be joined (cloning site). Moreover, the sites should lie on the regions of plasmid DNA which are not essential for their autonomous replication.

- They should carry as many gene codes as possible for phenotypically selectable markers to allow the detection of transformants. Moreover, if a foreign DNA fragment is joined into a cloning site, one of these selectable markers should be inactivated so as to identify the DNA insertion recombinants. An example of such a detection system for pBR322 was mentioned in (Section 4.1.4).
- They should be introduced into and remain stable in a wide variety of host bacterial strains.
- The cloned gene should be expressed at significant levels by transcription from the promotors present on the plasmid vector to allow phenotypical detection of foreign genes.

One of the most sophisticated groups of plasmid vectors for *E. coli* is the pUC plasmid group. The structure of a typical plasmid pUC19, is shown in Figure 4.6 (Vieira and Messing, 1982).

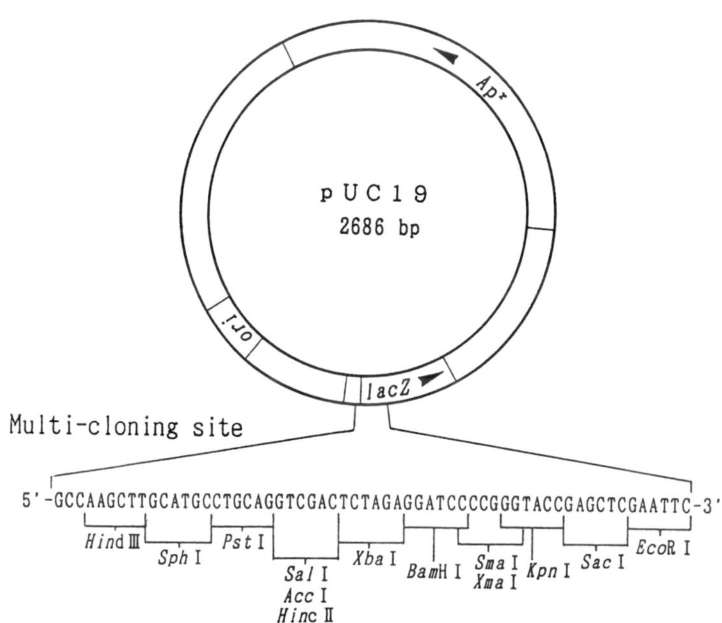

FIGURE 4.6. Structure of pUC19.

This vector is a small plasmid with a size of 2,686 base pairs of DNA and replicates under relaxed control, thereby existing in high copy numbers 50–60 copies per cell) in a host strain. Recognition sites for various restriction enzymes, each of which cleaves the plasmid at a single position, are contained in the *lacZ* region (multicloning sites) as follows: *Hind*III, *Sph*I, *Pst*I, *Sal*I (*Acc*I, *Hinc*II), *Xba*I, *Bam*HI, *Sma*I (*Xma*I), *Kpn*I, *Sac*I, and *Eco*RI. A selectable marker for pUC19 is resistance to ampicillin coded by *bla* gene. Cloned genes may be expressed by transcription from the promotor of the lactose operon of *E. coli* (*lac* promotor; *lac p/o*). The expression can be enhanced by induction with isopropyl-β-D-thiogalactopyranoside (IPTG). Moreover, direct screening of DNA insertion into multicloning sites can be carried out on media containing IPTG and 5-bromo-4-chloro-3-indolyl-β-D-garactopyranoside(X-gal), when *E. coli* JM103, *E. coli* JM105, etc. are used as host strains. This vector plasmid codes for α-peptide (one of the two components) of β-galactosidase (*lacZα* gene). When it is introduced into a host strain, such as *E. coli* JM103—which carries *lacZβ* gene coding for β-peptide of β-galactosidase but lacks *lacZα*—α-complementation of β-galactosidase occurs, resulting in the appearance of dark blue colonies of the metabolite of X-gal catalyzed by β-galactosidase. However, if the foreign DNA fragments are inserted into the multicloning sites which lie in the region including *lacZα*, β-galactosidase cannot be constructed in a host. Consequently, the transformants carrying pUC19 with a DNA insertion form white colonies on the media containing X-gal.

A few disadvantages of this vector must be noted. This plasmid has an extremely narrow host range; it cannot be introduced into bacterial hosts other than *E. coli*. Also, large DNA fragments cannot be joined efficiently to this vector.

Although various DNA manipulations of plasmids have been carried out to develop more versatile and useful cloning vectors, no others are as precise as pUC19. Researchers are seeking plasmid vectors to help solve even subtle problems.

For example, hybrid plasmids (combined plasmids) were constructed from two or more plasmids harbored by different bacterial species to extend the range of host vectors; such hybrid plasmids can be transformed into and maintained in every parent bacteria (Ehrlich, 1978).

Cosmid vectors were developed as hybrids of bacteriophage λ and plasmids in order to clone relatively large DNA fragments; the *cos* coding region of λ DNA was inserted into plasmid vectors. For example, a cosmid, pVK100, which was used in a cloning study described in Section 4.3, can accept a foreign DNA fragment with a size from 15.5–29.0 kb as an insert (Knauf and Nester, 1982). Transforming host cells with a large recombinant plasmid constructed with a cosmid is relatively simple, since the plasmid can be packaged into λ and thus introduced into an *E. coli* host at a high rate by transduction.

A mobile plasmid vector can be introduced from a donor bacteria to a secondary host by triparental conjugational transfer with the mediation of a helper plasmid such as RP4. The previously mentioned pVK100 can be transferred by triparental conjugation. Mobile plasmid vectors were constructed by the addition of the DNA fragments containing *mob* (plasmid mobilization), *ori* (the origins of replication), and *nic* (relaxation nick site) to plasmids. Plasmids carrying *mob, ori,* and *nic* can be mobilized by the mediation of the *tra* (transfer) gene of a helper plasmid.

Other improvements of plasmid vectors include the addition of selectable markers, available cloning sites, efficient transcription or controllable promotors, etc.

4.2.2 Broad Host Range Vectors for Gram-Negative Bacteria

Plasmid vectors for *Pseudomonas* strains are well-developed as are those for *E. coli* hosts. Most of these plasmid vectors can be transferred not only among *Pseudomonas* species but also into other gram negativa bacteria. They may be useful in the molecular-level breeding of a wide variety of wastewater treatment organisms, especially gram-negative bacteria, which generally dominate the processes (see Chapter 2). A gene cloned from a gram-negative soil or water bacterium (other than enteric bacteria, which are closely related to *E. coli*) often cannot be expressed in a *E. coli* host with a corresponding vector. To be fully expressed, these genes must be introduced into various gram-negative strains with broad host range vectors.

Broad host range plasmid vectors have been developed from naturally occurring plasmids as basic replicons. Some examples of naturally occurring broad host range plasmids and their characteristics are shown in Table 4.2. Among these plasmids, mainly RSF1010, Sa, and RK2 have been used as

TABLE 4.2. Broad host range plasmids.

Plasmid	Inc. Group	Size (kb)	Phenotype/Genotype
RK2, RP1, RP4, R68	P-1	60	*Ap, Km, Tc, tra*
RSF1010, R300B, R1162	Q	8.9	*Su, Sm, mob*
Sa	W	29.6	*Km, Cm, Sp, Su, tra*
pJP4	P-1	52	*Hg*, 2,4-D
pWWO	P-9	117	TOL/XYL

2,4-D: 2,4-dichlorophenoxyacetic acid degradation.
TOL/XYL: toluene/xylene degradation.
Other abbreviations are shown in the text.

replicons to develop useful broad host range vectors for gram-negative bacteria through time-consuming and burdensome DNA analysis and manipulations. Processes for constructing vectors use very similar methodology, regardless of the basic replicons used.

RSF1010, belonging to incompatibility group Q, is a nonconjugational, multicopy replicon with a size of 8.9 kb, and identical to R300B and R1162. It is smaller than the other broad host range plasmids listed in Table 4.2. Its structure (physical and functional map) is shown in Figure 4.7. Resistance to streptomycin (*Sm*) and sulphonamides (*Su*) and the regions necessary for plasmid mobilization (*ori, nic,* and *mob*) are coded on RSF1010. There are also three replication proteins (*rep*A, *rep*B, and *rep*C) which are essential for its replication.

Although RSF1010 has several recognition sites for restriction endonucleases, as shown in Figure 4.7, there are few desirable cloning sites, which may be a common property of broad host range plasmids. Among the restriction endonucleases which cleave RSF1010 at a single recognition site, *Hpa*I and *Pvu*II make blunt ends. Therefore, *Bst*EII, *Eco*RI, *Sst*I(*Sac*I) (along with *Pst*I, which has two adjacent sites) seem to be suitable for gene cloning. However, *Bst*EII cannot be used because DNA insertion into the recognition sites inactivates *Sm, Su,* or both "only-two" selectable markers. Unfortunately, *Sm* and *Su* are transcribed from the same promotor which lies in the region between two adjacent *Pst*I sites. Accordingly, when a foreign DNA fragment is inserted between the adjacent *Pst* I sites, both *Sm*

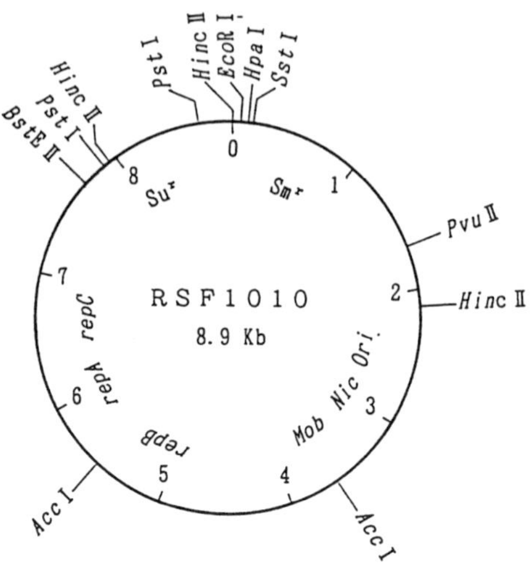

FIGURE 4.7. Structure of RSF1010.

and *Su* are inactivated because of the loss of the promotor sequence. On the other hand, since the *Sst* I site is included in the structural gene of *Sm,* insertion of a DNA fragment in it causes the loss of the *Sm* marker. Cloning in the *Eco*RI site, which does not exist in the structural gene, may also inactivate *Sm,* unless the inserted DNA fragment provides a new promotor (originating from a foreign DNA) for the expression of *Sm*. The inactivation of selectable markers and the fact that *Su* is not ideal for screening recombinants, led to the conclusion that only *Bst*EII is available as a cloning vector with RSF1010, and that *Pst* I, *Sst* I, and *Eco*RI cannot be used without other valid selectable markers. Therefore, more available cloning sites or selectable markers—which are not inactivated by the insertion of foreign DNA fragments—had to be added to RSF1010 to develop a useful broad host range vector.

Pursuing this strategy, samples of RSF1010-derived vectors, pKT210, pKT215, pKT248, pKT230, pKT231, and pKT247 were constructed as shown in Figure 4.8 (Bagdasarian et al., 1981). The small *Pst* I fragment (the region between the adjacent sites) of RSF1010 was replaced with DNA fragments of other plasmids, which are expressed as bold circular lines in the figure.

The pKT210 was constructed by replacing the 3.5-kb *Pst* I fragment of plasmid pSa which codes for resistance to chloramphenicol (*Cm* gene). With this modification, the *Eco*RI, *Sst* I, and *Hind* III sites can be used for gene cloning with pKT210. As the introduced pSa-derived fragment carried a *Bst* EII and a *Hind* III site, pKT210 has two recognition sites for *Bst* EII and a single site for *Hind* III. Moreover, the transformants carrying recombinant plasmids with inserted DNA fragments generated by digestion with *Eco*RI and *Sst* I can be identified by screening streptomycin sensitive phenotypes (*Cm* remaining). The pKT215 carries the same pSa-derived *Cm*-encoding DNA fragment, but in the reverse direction of pKT210.

Another derivative carrying the *Cm* marker, pKT248, was constructed by inserting 4.8-kb *Pst* I fragment from the R621ala plasmid, which shows resistance to ampicillin (*Ap*), *Cm,* kanamycin (*Km*), and tetracycline (*Tc*) (Hedges, 1974), into the *Pst* I site that is nearest to the *Bst* EII site of RSF1010. This vector contains available cloning sites for *Sal* I in the *Cm* region in addition to *Bst* EII, *Eco*RI and *Sst* I each of which cleaves at a single RSF1010 site.

On the other hand, pKT230 and pKT231 were constructed to obtain broad host range vectors showing *Sm* and *Km* as the derivatives of RSF1010. The pKT230 was generated by incorporating *Pst* I-digested RSF1010 and *Pst* I-linearized plasmid pACYC177 (Chang and Cohen, 1978). These double replicon vector codes for *Sm* and *Km* can accept a foreign DNA fragment into a single recognition site for *Hind* III, *Xma* I, *Xho* I, *Bam*HI, *Bst* EII,

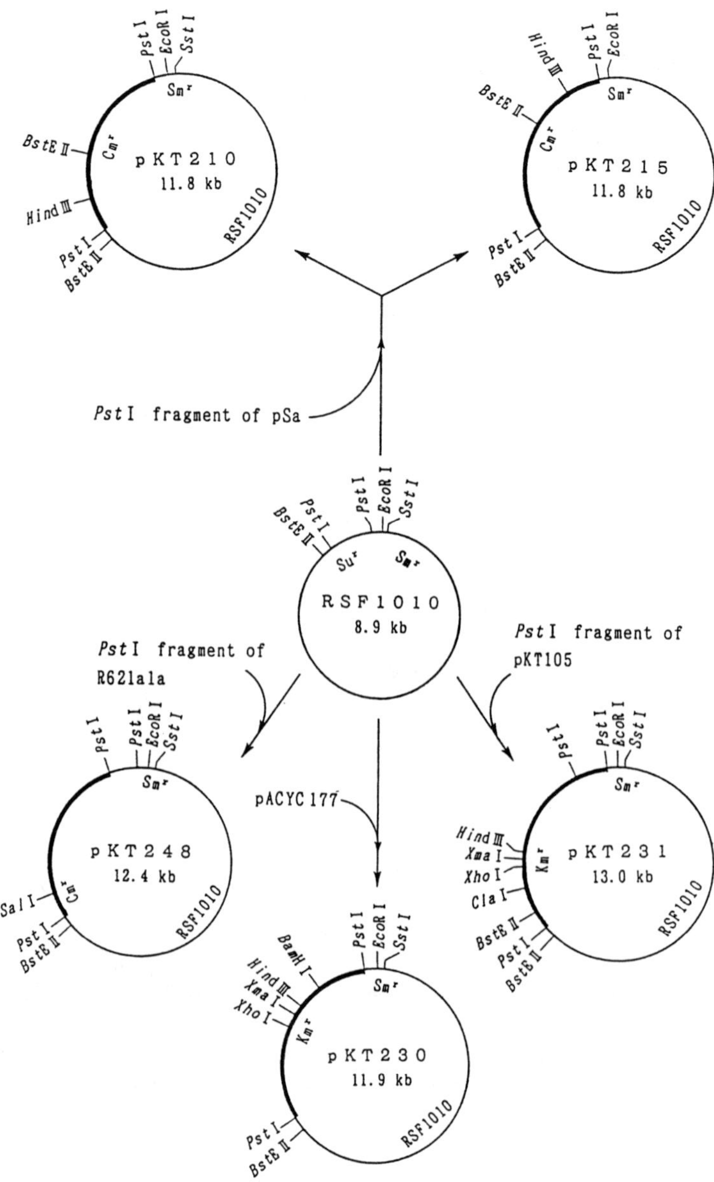

FIGURE 4.8. Construction of RSF1010-derived broad host range vectors.

*Eco*RI, or *Sst*I. The pKT231, which was constructed by the replacement of the small *Pst*I fragment of RSF1010 with two *Pst*I-generated (3.8-kb and 1.0-kb) DNA fragments of a miniplasmid derivative of R6-5 (pKT105), also encodes *Sm* and *Km,* and has cloning sites for the fragment digested with *Hind*III, *Xma*I, *Xho*I, *Cla*I, *Eco*RI, and *Sst*I.

Host ranges of such RSF1010-derived vectors were investigated by the research group of Bagdasarian and Timmis (Bagdasarian et al., 1981; Bagdasarian and Timmis, 1982). A series of RSF1010-derived vectors were introduced into and replicated in host strains of *E. coli, P. aeruginosa, P. putida, Agrobacterium tumefaciens, Azotobacter vinelandii, Alcaligenes faecalis, Alcaligenes eutroophus, Acetobacter xylinum, Metylophilus methylotrophus, Rhodopseudomonas spheroides, Rhizobium meliloti, Rhizobium leguminosarum, Caulobacter crescentus, Yersinia enterocolitica, Klebsiella aerogenes, Serratia marcescens, Erwinia coratovora, Xanthomonas compestris,* and *Gluconobacter* sp., suggesting that RSF1010 has the broadest range of host vectors useful for genetic manipulation. Most RSF1010-derived vectors have another advantage in that they can be mobilized by helper-mediated conjugation, as they contain the *ori, nic,* and *mob* regions of the original replicon. A disadvantage of RSF1010-derived vectors is that they are considerably larger than those developed for a narrow range of *E. coli*-related hosts such as pUC19. This disadvantage arises from the fact that the genes for plasmid replication (*repA, repB, repC,* and *ori*) straddle a large region (approximately 5 kb) of RSF1010.

The RSF1010 replicon allow it to be used to construct other versatile broad host range plasmid vectors. For example, broad host range cosmid vectors—which can be utilized for cloning very large DNA fragments, packaged into λ phage particles, and infected into *E. coli* strains that may serve as intermediate hosts prior to the transfer of recombinant cosmids into other gram negative bacterial hosts—include pKT247 (Bagdasarian et al., 1981), pMMB33, and pMMB34 (Frey et al., 1983); broad host range regulated expression vectors, which contain a *tac* promotor that causes abundant expression of cloned genes when induced with IPTG, pMMB22 and pMMB24 (deBoer et al., 1983; Bagdasarian et al., 1984); vectors defective in mobile properties (*mob*$^-$ derivatives), constructed with consideration for government regulations concerning recombinant DNA experiments, pKT261, pKT262, pKT263, and pKT264 (Bagdasarian et al., 1981); and a promotor probe vector, pKT240, which can be used to identify the promotor regions and terminators of cloned DNA fragments (Franklin, 1985).

Similar general or special purpose broad host range vectors have been developed on the basis of pSa and RK2 replicons. The pSa is an incompatibility group W plasmid with a size of 29.6 kb, and which specifies antibiotic resistances (*Km, Cm, Su,* and spectinomycin). RK2, on the other hand, is a 60-kb, *Ap-*, *Km-*, and *Tc*-encoding plasmid belonging to incompatibility group P-1. Both plasmids are considerably larger than RSF1010, and both are self-transmissible (coding for *tra*).

In the construction of pSa- or RK2-derived broad host range vectors, small plasmid derivatives, such as *mini-Sa,* were generated by first deleting the regions coding for conjugal transfer (*tra*), then introducing further modifications, such as selectable markers or unique restriction endonuclease recognition sites, to generate available cloning sites. The pSa-derived vectors include pSa4 (Tait et al., 1983), pGV1106, and pGV1122 (Leemans et al., 1982). On the other hand, the RK2-derived vectors are pRK290, pRK2501 (Haas, 1983), and a cosmid, pLAFR1 (Friedman et al., 1982).

Since RSF1010, pSa, and RK2 belong to different incompatibility groups, vectors and recombinant plasmids based on these vectors can coexist in the same host cell. This allows the complementation analysis of more than two cloned genes.

4.2.3 Vectors for Gram-Positive Bacteria

Gram-positive bacteria identified mainly as *Bacillus* sp., coryneform group, etc., have been isolated at relatively low frequencies from wastewater treatment microcosms (see Section 2.3). Vectors which can introduce cloned genes into gram-positive bacteria may be useful in some cases for GEM construction in wastewater treatment.

Vectors for gram-positive bacteria are not as well-developed as they are for *E. coli* and *Pseudomonas* (broad host range). However, a relatively large number of plasmid vectors have been constructed for *Bacillus* sp., since the hosts are able to secrete extracellular protein; secreted proteins can be recovered and purified more easily than intercellular proteins. In the *Bacillus* cloning systems, commonly used host strains is *B. subtilis.* Cloning vectors have been constructed for naturally occurring replicons which can replicate in *B. subtilis.* Naturally occurring plasmids and derivatives (vectors) which can be used for gene cloning and expression in *Bacillus* sp. are shown in Table 4.3 (Hardy, 1985).

As shown in the table, many vectors were developed on the basis of replicons originally found in *Staphylococcus aureus,* such as pUB110. The pUB110 has many homologous regions with pBC16 which originated from *Bacillus cereus,* and is a representative *Bacillus* plasmid; this may allow its replication in *Bacillus* hosts. Interestingly, many naturally occurring plasmids detected in gram-positive bacteria were very similar to each other, suggesting that these plasmids or vectors may have a wide range of gram-

TABLE 4.3. Plasmid vectors for Bacillus.*

Plasmid	Source	Markers
pBC16	*Bacillus cereus*	*Tc*, naturally occurring
pAB124	*B. stearothermophilus*	*Tc*, naturally occurring
pUB110	*Staphylococcus aureus*	*Km*, naturally occurring
pSA501	*S. aureus*	*Sm*, naturally occurring
pSA2100	*S. aureus*	*Cm, Sm*, naturally occurring
pC194	*S. aureus*	*Cm*, naturally occurring
pBD6	pSA501, pUB110	*Sm, Km*, vector
pBD8	pSA2100, pUB110	*Sm, Km, Cm*, vector
pBD64	pC194, pUB110	*Km, Cm*, vector
pHV41	pC194, pUB110, pBR322	*Km, Cm*, shuttle vector

*Referring to Hardy (Hardy, 1985).

positive bacteria hosts. The physical map of pUB110 is shown in Figure 4.9. This replicon is 4.5 kb in size, and specifies *Km* which straddles the single recognition site for *Bgl*II. The development of vectors for *Bacillus* sp. from such replicons was methodologically similar to that for *E. coli* and gram-negative bacteria.

Moreover, "shuttle" vectors, which can replicate in both *B. subtilis* and *E. coli*, have been generated by fusion of replicons derived from both parent-related bacteria and minor modifications. They are listed in Table 4.3. Such vectors are used to introduce genes which were already cloned using *E. coli* or *Pseudomonas* systems into gram-positive *Bacillus subtilis* strains.

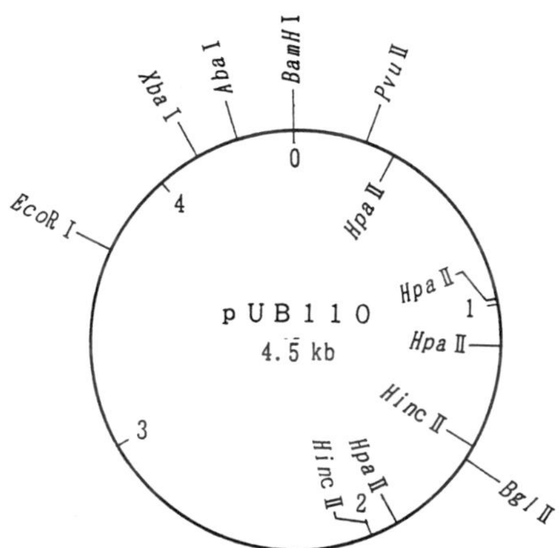

FIGURE 4.9. Structure of pUB110.

Another putatively important gram-positive bacterial group for wastewater treatment, coryneform bacteria, have no useful gene cloning systems to date. However, cryptic small plasmids were isolated from the representative genera *Arthrobacter* strains, and useful vectors for them are now being developed (Plakidou-Dymock et al., 1991).

4.3 CLONING OF PHENOL CATABOLIC GENES

4.3.1 Phenol Catabolic Pathway of *P. putida* BH

Phenol is a troublesome xenobiotic which is toxic to wastewater treatment organisms. It appeared that phenol catabolic genes might be a useful material for breeding wastewater treatment GEMs.

We isolated a phenol degrading bacterium, *P. putida* BH, from activated sludge which was acclimated to phenol. This strain can efficiently metabolize phenol and its methylated derivatives, *m*-, *p*-, and *o*-cresol, and 3,4-dimethylphenol via the *meta* cleavage pathway of catechol (or methylated catechol). The catabolic pathway of phenols in *P. putida* BH is shown in Figure 4.10. The figure also includes the corresponding catabolic enzymes and genes (*phe* genes) which we had presumed. In addition to phenols, *P. putida* BH can utilize a wide variety of aromatic compounds as the sole source of carbon and energy, including benzoate, protocatechuate, *p*-hydroxybenzoate, and *m*-toluate. But when these aromatic compounds are degraded, C230—which is the key enzyme of the *meta* cleavage pathway—is not induced, suggesting the metabolism through the *ortho* cleavage pathway of catechol or protocatechuate. On the other hand, toluene, xylene, benzene, aminophenols, nitrophenols, phthalate, phenoxyacetate cannot support the growth of this strain.

To date, a few different genes specifying phenol hydroxylase and enzymes responsible for catechol produced from phenol have been cloned and analyzed genetically (Kukor and Olsen, 1990; 1991; Kivisaar et al., 1990; 1991; Nurk et al., 1991; Bartilson et al., 1990; Nordland et al., 1990; Shingler et al., 1992).

The *tbu* genes encoding phenol catabolism via the *meta* cleavage pathway were isolated from the chromosomal DNA of *Pseudomonas pikettii* PK01 which utilizes benzene and toluene via phenolic intermediates (phenol and *m*-cresol). The phenol hydroxylase-encoding *tbuD* was approximately 2.0 kb in size. Catechol catabolic genes (*tbuEFGHIJK*) were regulated as the operon other than *tbuD* (Kukor and Olsen, 1990; 1991).

The *dmp* genes coding for complete catabolism of phenol/3,4-dimethylphenol via the *meta* pathway in *Pseudomonas* sp. CF600 were cloned

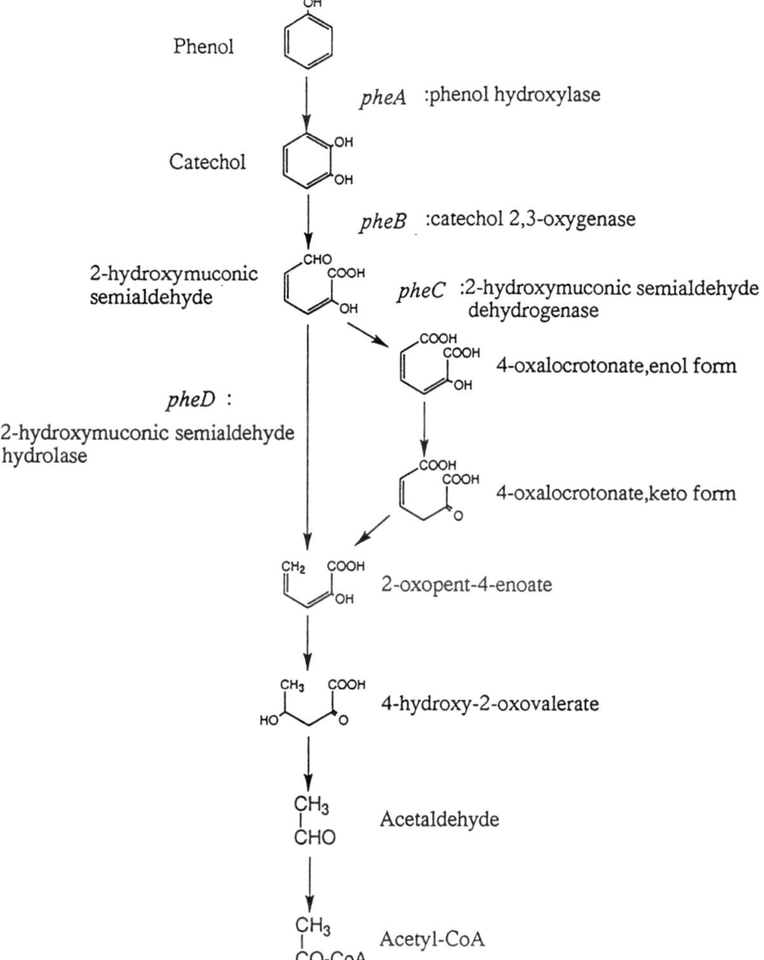

FIGURE 4.10. Phenol catabolic pathway in *P. putida* BH. Presumed corresponding catabolic genes *pheA–pheD* and their coding enzymes are shown.

from a large plasmid pVI150 belonging to incompatibility group P2. The phenol hydroxylase coded by *dmpA* was composed of six components of polypeptide, and the *dmpA* (*dmpKLMNOP*) was approximately 5.5 kb in size (Bartilson et al., 1990; Nordland et al., 1990; Shingler et al., 1992).

Since no plasmids could be detected in *P. putida* BH when screened by methods described in Chapter 2, we tried to clone the phenol catabolic genes from the chromosomal DNA. Genes for the entire phenol catabolic pathway, those necessary for the genetic analysis, are considered to be clustered. For example, the genes coding for the phenol/3,4-dimethylphenol pathway of *Pseudomonas* sp. CF600 were found clustered on the approximately 13-kb region of plasmid pVI150 as a single operon structure (Shingler et al., 1992). Therefore, the broad host range cosmid vector pVK100 was used as a cloning vector to clone a rather large DNA fragment.

4.3.2 Cosmid Cloning

A physical and functional map of pVK100 used for the cloning of *phe* genes is shown in Figure 4.11. The vector maintained in an *E. coli* host was extracted by the Birmboin and Doly method (Birmboin and Doly, 1979), and purified further by cesium chloride gradient centrifugation (Dazin and Chakrabarty, 1984). Total genomic DNA of *P. putida* BH was prepared by using the freeze-thaw lysis cycle and the phenol extraction method described by Saito and Miura (1963).

Packaging the recombinant DNA into λ phage particles selects for DNA molecules with a size of 38.5–52.0 kb, which is the size of the cosmid plus

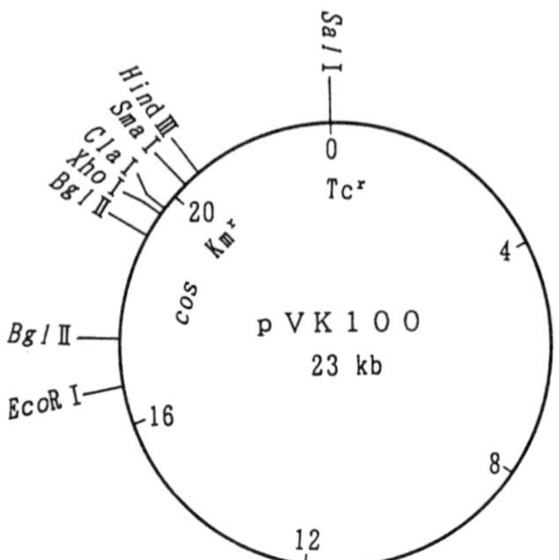

FIGURE 4.11. Structure of pVK100.

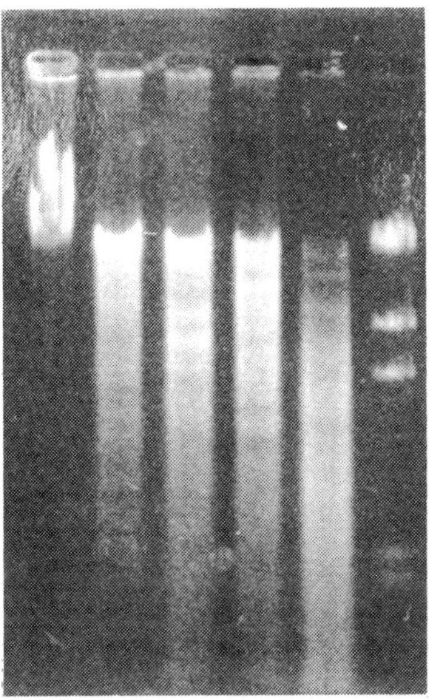

FIGURE 4.12. Partial digestion of chromosomal DNA of *P. putida* BH: lane A, chromosomal DNA of *P. putida* BH; lanes B, C, D, E, partially digested chromosomal DNA; lane H, *Hind* III-digested λ DNA.

the foreign DNA fragment. Therefore, when using 23.0 kb pVK100 as a cloning vector, only DNA fragments from 15.5–29.0 kb can be cloned successfully. The genomic DNA of *P. putida* BH was partially digested with *Sal*I to produce such DNA fragments efficiently. Complete digestion with *Sal*I causes the generation of much smaller DNA fragments. Partial digestion of DNA was performed by controlling either the total amount of the restriction enzyme used, or the reaction time. The size fractionation of partially digested DNA was performed by agarose gel electrophoresis (Figure 4.12), and the DNA fragments with appropriate sizes were collected by electroelution from a piece of gel containing such fractions.

Then, partially digested genomic DNA and pVK100, which was completely digested with *Sal*I, were ligated with T4 DNA ligase. During ligation, the foreign DNA fragment must be linked to cosmids to generate concatenated molecules in which the *cos* sites of the two vectors flank a foreign DNA fragment, and in which the two *cos* sites are in the same orientation, indicating that the entire complement of plasmid information is contained between the two *cos* sites (see Figure 4.5). Only such concatenated DNA molecules can be packaged into λ phage particles; the *cos* sites flanking the foreign DNA are cleaved and the intervening DNA is packaged into λ heads. To prevent vector-to-vector ligation, we treated the linearized pVK100 with alkaline phosphotase prior to ligation. Finally, the ligated DNA molecules were packaged into λ phage head (Amersham) according to the supplier's specifications, and then transfected into *E. coli* HB101.

About 20,000 transductants appeared on the selective medium containing kanamycin. Fifteen colonies (clones) were randomly selected and their cosmids were analyzed. Since fourteen of them had an average of 20-kb DNA insert from *P. putida* BH, the genomic library constructed was adequate to include the whole chromosome of the DNA donor.

More than 2,300 transformants of *E. coli* HB101 showing *Km*(i.e., containing chimeric cosmids or cosmid-dimer) were transferred onto the selective medium containing phenol for identifying clones expressing phenol hydroxylase. A total of seven presumably positive clones were obtained by this phenotypic screening—colonies of the two clones accumulated a brownish compound, which seemed to be produced by the spontaneous oxidation of catechol on the medium after incubation at 37°C for 2 days, suggesting that they harbored the phenol hydroxylase gene (*pheA*) but no other genes coding for *meta* cleavage pathway of catechol. Colonies of the other five clones turned a yellowish color, which seemed due to the accumulation of 2-HMS, then the color disappeared, suggesting they contained *pheA*, C230 gene (*pheB*), and genes for further degradation, at least one that codes for the enzyme catalyzing the transformation of 2-HMS into the colorless compound (4-oxalocrotonate or 2-oxopent-4-enoate). These seven clones which were positive for at least *pheA* and their chimeric cosmids are shown in Table 4.4.

Recombinant cosmids shown in Table 4.4 were analyzed by *Sal*I digestion, resulting in five patterns of inserted DNA fragments, which are shown in Figure 4.13. The cosmids pS4-92 and pS18-93, and pS3-43 and pS23-97 were identical, respectively. From the analysis of *Sal*I-digested DNA frag-

TABLE 4.4. Positive clones for phenol catabolic genes.

Clone	Plasmid	Insert DNA Fragment	*pheA*	*pheB*
S0492	pS4-92	22.6-kb *Sal*I fragment	+	–
S1893	p18-93	22.6-kb *Sal*I fragment	+	–
S0999	pS9-99	27.8-kb *Sal*I fragment	+	–
S0343	pS3-43	23.4-kb *Sal*I fragment	+	+
S2397	pS23-97	23.4-kb *Sal*I fragment	+	+
S1921	pS19-21	15.5-kb *Sal*I fragment	+	+
S1045	pS10-45	19.6-kb *Sal*I fragment	+	+

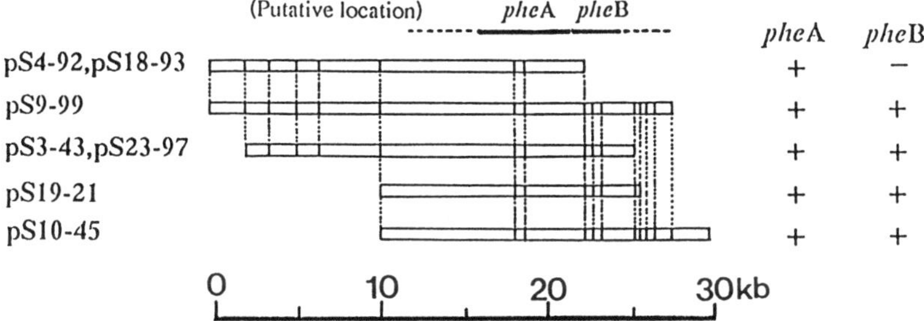

FIGURE 4.13. Recombinant cosmids positive for phenol catabolic genes. Dotted lines indicate *Sal*I sites for inserted DNA fragments of recombinant cosmids containing phenol catabolic genes. The presence of *pheA* and *pheB* were determined by the production or accumulation of catechol and 2-HMS on LB agar supplemented with phenol.

ments and the above mentioned coloration (phenotypic detection of phenol degrading genes in recombinants), the approximate locations of *pheA* and *pheB* genes are also described in the figure.

4.3.3 Expression of Cloned Phenol Hydroxylase Gene

None of the seven positive clones detected by coloration on the selective medium could grow on phenol as a sole source of carbon and energy. This suggested either that recombinant cosmids obtained did not contain whole genes essential for complete phenol degradation, or that phenol degrading genes could not express in *E. coli* host at sufficiently high levels to support growth on phenol. Thus, although these clones were grown in LB broth containing phenol at the concentration of 400 mg/L, and the phenol concentration was monitored by HPLC analysis, no detectable breakdown of phenol was observed, indicating little or extremely low expression of the *pheA* gene in *E. coli* HB101.

From five distinct recombinant cosmids, pS10-45 was used to further study the expression of cloned *pheA* gene. To compare the expression of *pheA* in *E. coli* with that in *Pseudomonas* host, the pS10-45 was introduced into *P. putida* KT2440, which lacks phenol hydroxylase but does carry the enzymes for the *ortho*-cleavage pathway of catechol (by triparental conjugation with a helper plasmid, pRK2013). The resultant transconjugant *P. putida* KT2440 (pS10-45) was able to utilize and grow on phenol as a sole carbon and energy source.

The expression of *pheA* in *E. coli* HB101 and *P. putida* KT2440 was assayed by measuring oxygen uptake of intact cells. Each recombinant harboring pS10-45 precultured in LB broth was inoculated into 100 mL of minimal salts medium containing 0.2 g/L yeast extract, 0.1 g/L casamino acid, 11 mM sodium acetate in the presence or absence of phenol until midlog phase at 37 °C for *E. coli* or at 30 °C for *P. putida*, respectively. Cells were harvested by centrifugation and washed twice with 100 mM phosphate buffer. Then the cells were suspended in the same buffer at a concentration of OD_{600} of about 0.5, and the suspension was kept at 4 °C until it was used for the measurement of oxygen uptake within 2 hours. Oxygen uptake was monitored with a Clark-type oxygen electrode at 30 °C in a 6-mL chamber. The reaction mixture, with a final volume of 6 mL, consisted of 0.18 mL of 83.3 mM phenol solution and 5.72 mL of the cell suspension.

Results are summarized in Table 4.5. As shown in the table, although *P. putida* KT2440 (pS10-45) expressed *pheA* at a high level when induced with phenol, the oxygen uptake rate of *E. coli* C600 (pS10-45) was at an undetectable level, notwithstanding that catechol production was detected as coloration on the test plates. As the *pheA* on pS10-45 expressed inducibility in *P. putida* KT2440, the regulatory system for phenol degradation may be present on the insert fragment of pS10-45.

4.3.4 Analysis of Cloned Fragments

To localize the *pheA* gene encoding phenol hydroxylase, which is the most important enzyme on the phenol catabolic pathway, a series of DNA fragments from pS10-45 were subcloned in *E. coli* JM103 using a vector pUC19, which has multiple cloning sites for a wide variety of restriction endonucleases. Figure 4.14 shows the resultant plasmids containing various DNA fragments from the insert and the restriction map of pS10-45.

The expression of *pheA* in *E. coli* JM103 harboring such a series of plasmids was investigated on LB medium plate containing phenol, kanamy-

TABLE 4.5. Expression of pheA *in* E. coli *and* P. putida *host.*

Strain	Phenol Hydroxylase Activity	
	Phenol-Induced Cell	Not Induced Cell
P. putida		
BH	648	<20
KT2440	<20	<20
KT2440 (pS10-45)	230	<20
E. coli		
HB101	<20	<20
HB101 (pS10-45)	<20	<20

Phenol hydroxylase activity is expressed as μmol-O_2 consumption/min/g-cell.

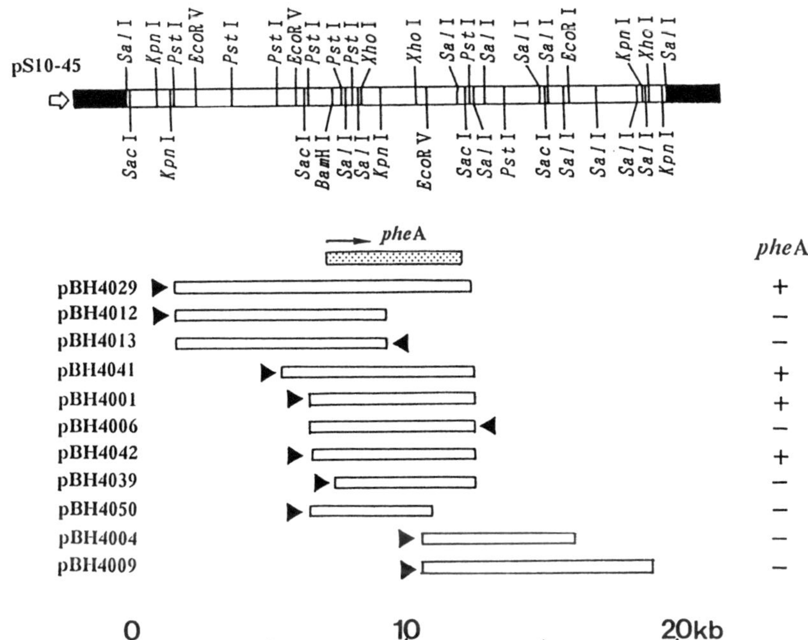

FIGURE 4.14. Restriction map of the insert DNA fragment of pS10-45 and subcloning of *pheA:* ▭, BH chromosomal DNA; ■, pVK100; ⇨, *Km* promotor on pVK100; ▶, *lac* promotor on pUC19.

cin, and IPTG. The recombinants harboring pBH4029, pBH4041, pBH4001, and pBH4042 exhibited brownish color on the test plates due to catechol accumulation and its spontaneous oxidation. Amongst such plasmids testing positive for catechol accumulation, pBH4042 had the smallest DNA insert fragment (5.7-kb *Pst*I-*Sac* fragment), which allowed the host strain to express the *pheA* gene. However, pBH4039 and pBH4050 did not confer catechol production ability upon the host strain. From these results, the approximate location of the *pheA* gene on pS10-45 was estimated as shown in Figure 4.14.

Although the recombinant harboring pBH4001 could produce catechol from phenol, the recombinant harboring pBH4006 containing the same DNA fragment in the opposite direction to the vector promotor could not. Therefore, it is thought that the transcription of *pheA* may depend on the *lac* promotor carried on pUC19, and that its transcriptional direction is from left to right in Figure 4.14.

As mentioned in Section 4.3.3, it was obvious that in the regulatory gene controlling the phenol degradation genes, at least the *pheA* gene was cloned on pS10-45. Therefore, the complementation tests of various DNA fragments from the *P. putida* BH DNA insert of pS10-45 with the *pheA* were performed as follows in order to identify the region enhancing the expression of *pheA*. For the tests, compatible vectors pUC19 and pKT230 were used.

First, the 6.0-kb *Sac*I-*Sac*I fragment, which allowed the host *E. coli* strain to express *pheA*, was inserted into the *Sac*I site of pKT230, and the resultant plasmid pBH4201 was constructed. *E. coli* JM103 harboring pBH4201 showed faint brownish color on LB medium containing phenol after 2-day incubation at 37°C due to the expression of *pheA* at a low level by transcription from the *Km* promotor on the vector pKT230. Subsequently, various DNA fragments of pS10-45 were subcloned into pUC19, and resultant plasmids were introduced into *E. coli* JM103 containing pBH4201 for complementation.

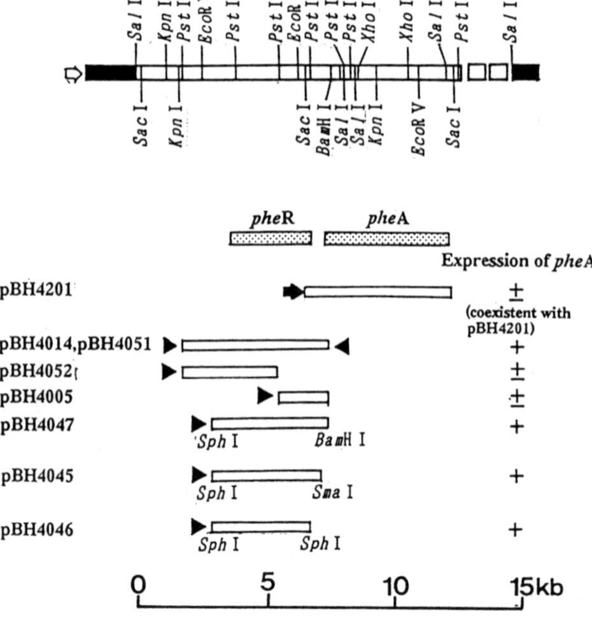

FIGURE 4.15. Subcloning of *pheR*. Subcloned fragments in pUC19 were introduced to complement *pheA*-containing pBH4201. Putative location of *pheR* was determined by the enhancement of catechol accumulation (+) on LB agar supplemented with phenol compared with JM103 (pBH4201) (±). ▭, BH chromosomal DNA; ■, pVK100; ➡, *Km* promotor on pKT230; ▶, *lac* promotor on pUC19.

TABLE 4.6. Complementation test of pheA *and* pheR *in* E. coli *JM103.*

Plasmid	Phenol Hydroxylase Activity	
	Phenol-Induced Cell	Not Induced Cell
pS4-92 (cosmid)	5.05	<0.1
pBH4201	0.80	0.90
pBH4045	<0.1	<0.1
pBH4045, pBH4201	7.23	8.17
pBH4046, pBH4201	4.07	n.t.
pBH4047, pBH4201	10.26	n.t.

Phenol hydroxylase activity is expressed as μmol-catechol produced/hr/g-cell.
n.t.: not tested.

A series of recombinants harboring both plasmids was tested for catechol-producing ability on LB medium containing phenol. It was found that some plasmids containing the upper DNA region just adjacent to the approximate *pheA* region—for example, pBH4045, pBH4046, and pBH4047—complemented pBH4201 to enhance the expression of *pheA* in *E. coli* JM103. The brownish color on the test medium exhibited by *E. coli* JM103 harboring both plasmids (pBH4201 and pBH4045) was much darker than that of the recombinant harboring only pBH4201 (Figure 4.15). Thus, these plasmids are thought to contain the positive regulatory gene (*pheR*) for the phenol degradation genes (at least *pheA*). In Figure 4.15, the putative location of *pheR* is also shown.

The enhancement of the *pheA* expression coded on pBH4201 by the complementation of the above-mentioned plasmids was quantitatively evaluated as follows. The recombinants were grown in LB broth with or without phenol and the washed cells were prepared as described in Section 4.3.3 for oxygen uptake assay, except that they were suspended at 1.0 of OD_{600} in minimal salts medium containing phenol. Phenol degradation was started by shaking the cell suspension at 30°C in a 300-mL flask (working volume of 100 mL) on a reciprocal shaker (160 rpm), and the catechol accumulation rate, i.e., phenol hydroxylase activity, for each strain was measured by a sensitive catechol determination method developed in our laboratory (Fujita et al., 1991). Results are shown in Table 4.6.

It is apparent that the catechol accumulation or production rate of *E. coli* JM103 (pBH4201) was enhanced by the complementation of pBH4045, pBH4046, and pBH4047. When pBH4045 coexisted with pBH4201 in *E. coli* JM103, the catechol accumulation rate observed was more than 50 times higher than that observed without pBH4045. Interestingly, the *pheA* gene expressed in *E. coli* JM103 (pBH4201/pBH4045) suggests that *pheR* may not require phenol as a cofactor for inducing phenol catabolic enzymes.

Further investigation into the phenol catabolic genes and their regulatory system(s) is going on now.

4.4 REFERENCES

Bagdasarian, M. M., E. Amann, R. Lurz, B. Rückert and M. Bagdasarian. 1984. "Activity of the Hybrid *trp-lac* (tac) Promotor of *Escherichia coli* in *Pseudomonas putida*. Construction of Broad-Host-Range, Controlled-Expression Vectors," *Gene,* 26:273–282.

Bagdasarian, M., R. Luze, B. Rückert, F. C. H. Franklin, M. M. Bagdasarian, J. Frey and K. N. Timmis. 1981. "Specific-Purpose Plasmid Cloning Vectors, II. Broad Host Range, High Copy Number, RSF1010-Derived Vectors and Host-Vector System for Gene Cloning in *Pseudomonas,*" *Gene,* 16:237–247.

Bagdasarian, M. and K. N. Timmis. 1982. "Host/Vector Systems for Gene Cloning in *Pseudomonas,*" *Curr. Topics Microbiol. Immunol.*, 96:47–67.

Bartilson, M., I. Nordlund and V. Shingler. 1990. "Location and Organization of the Dimethylphenol Catabolic Genes of *Pseudomonas* CF600," *Mol. Gen. Genet.*, 220:294–300.

Birnboim, H. C. and J. Doly. 1979. "A Rapid Alkaline Extraction Procedure for Screening Recombinant Plasmid DNA," *Nucleic Acid Res.*, 7:1513–1523.

Bolivar, F., R. L. Rodrigez, P. J. Greene, M. C. Betlach, H. L. Heyneker and H. W. Boyer. 1977. "Construction and Characterization of New Cloning Vehicles II. A Multi Purpose Cloning System (Recombinant DNA; Molecular Cloning Plasmid Vector; EK2 Host-Vector System)," *Gene,* 2:95–113.

Chakrabarty, A. M., ed. 1978. *Genetic Engineering.* New York: CRC Press, Inc.

Chakrabarty, A. M., J. R. Mylroie, D. A. Friello and J. G. Vacca. 1975. "Transformation of *Pseudomonas putida* and *Esherichia coli* with Plasmid-Linked Drug-Resistance Factor DNA," *Proc. Natl. Acad. Sci.*, 72:3647–3651.

Chang, A. C. Y. and S. N. Cohen. 1978. "Construction and Characterization of Amplifiable Multicopy DNA Cloning Vehicles Derived from the p15A Cryptic Miniplasmid," *J. Bacteriol.*, 134:1141–1156.

Clark, L. and J. Carbon. 1976. "A Colony Bank Containing Synthetic ColEI Hybrid Plasmids Representative of the Entire *E. coli* Gene," *Cell,* 9:91–99.

Dazin, A. and A. M. Chakrabarty. 1984. "Cloning of Genes Controlling Alginate Biosynthesis from Mucoid Cystic Fibrosis Isolate of *Pseudomonas aeruginosa,*" *J. Bacteriol.*, 159:9–18.

DeBoer, H. A., L. J. Comstock and M. Vasser. 1983. "The *tac* Promotor: A Functional Hybrid Derived from the *trp* and *lac* Promotors," *Proc. Natl. Acad. Sci.*, 80:21–25.

Ehrlich, S. D. 1978. "DNA Cloning in *Bacillus subtilis,*" *Proc. Natl. Acad. Sci.*, 75:1433–1436.

Franklin. 1985. "Cloning Vectors for Gram Negative Bacteria," in *DNA Cloning,* Oxford: JRL Press Ltd., pp. 165–184.

Frey, J., M. Bagdasarian, D. Feiss, F. C. H. Franklin and J. Deshusses. 1983. "Stable Cosmid Vectors That Enable the Introduction of Cloned Fragments into a Wide Range of Gram-Negative Bacteria," *Gene,* 24:299–308.

Friedman, A. M., S. R. Long, S. E. Brown, W. J. Buikema and F. M. Ausubel. 1982. "Construction of a Broad Host Range Cosmid Cloning Vector and Its Use in the Genetic Analysis of *Rhizobium mutans*," Gene, 18:289–296.

Fujita, M., T. Kamiya, M. Ike, Y. Kawagoshi and N. Shinohara. 1991. "Catechol 2,3-oxygenase Production by Genetically Engineered *Escherichia coli* and Its Application to Catechol Determination," *Wld. J. Microbiol. Biotechnol.*, 7:407–414.

Glover, D. M., ed. 1985. *DNA Cloning I, II.* Oxford: IRL Press Ltd.

Haas, D. 1983. "Genetic Aspects of Biodegradation by Pseudomonads," *Experimentia*, 39:1199–1213.

Hanahan, D. 1985. "Techniques for Transformation of *E. coli*," in *DNA Cloning I.* Oxford: IRL Press Ltd., pp. 109–135.

Hardy, K. G. 1985. "*Bacillus* Cloning Methods," in *DNA Cloning II.* Oxford: IRL Press Ltd., pp. 109–135.

Hedges, R. W. 1974. "*R* Factor from *Providencia*," *J. Gen. Microbiol.*, 81:171–181.

Holloway, B. W. and A. F. Morgan. 1986. "Genomic Organization in *Pseudomonas*," *Ann. Rev. Microbiol.*, 40:79–105.

Kivisaar, M. A., R. Horak, L. Kasak, A. L. Heinaru and J. K. Habicht. 1990. "Selection of Independent Plasmid Determining Phenol Degradation in *Pseudomonas putida* and the Cloning and Expression of Genes Encoding Phenol Monooxygenase and Catechol 1,2-dioxygenase," *Plasmid*, 24:25–36.

Kivisaar, M. A., A. Nurk and L. Kasak. 1991. "Sequence of the Plasmid-Encoded Catechol 1,2-Dioxygenase Gene, *PheB*, of Phenol-Degrading *Pseudomonas* sp. Strain EST1001," *Gene*, 98:15–20.

Knauf, V. C. and E. W. Nester. 1982. "Wide Host Range Cloning Vectors: Cosmid Clone Bank of *Agrobacterium* Ti Plasmids," *Plasmid*, 8:45–54.

Kukor, J. J. and R. H. Olsen. 1990. "Molecular Cloning, Characterization, and Regulation of a *Pseudomonas pikettii* PK01 Gene Encoding Phenol Hydroxylase and Expression of the Gene in *Pseudomonas aeruginosa* PA01c," *J. Bacteriol.*, 172:4624–4630.

Kukor, J. J. and R. H. Olsen. 1991. "Genetic Organization and Regulation of *Meta* Cleavage Pathway for Catechols Produced from Catabolism of Toluene, Benzene, and Cresol by *Pseudomonas pikettii* PK01," *J. Bacteriol.*, 173:4587–4594.

Leemans, J., J. Langenakens, H. DeGreve, R. Deblaere, M. Van Montagu and J. Schell. 1982. "Broad-Host-Range Cloning Vectors Derived from the *W*-Plasmid Sa," *Gene*, 19:361–364.

Maniatis, T., E. F. Fritsch and J. Shambrook. 1982. *Molecular Cloning–A Laboratory Manual.* Cold Spring Harbor, New York: Cold Spring Harbor Laboratory.

Murray, K. 1978. "Restriction Enzymes and Their Uses in Genetic Engineering," in *Genetic Engineering*, A. M. Chakrabarty, ed., New York: CRC Press, Inc., pp. 113–122.

Nordlund, I., J. Powlowski and V. Shingler. 1990. "Complete Nucleotide Sequence and Polypeptide Analysis of Multicomponent Phenol Hydroxylase from *Pseudomonas* sp. Strain CF600," *J. Bacteriol.*, 172:6826–6833.

Nurk, A., L. Kasak and M. A. Kivisaar. 1991. "Sequence of the Gene (*PheA*) Encoding Phenol Monoxygenase from *Pseudomonas* sp. Strain EST1001," *Gene*, 102:13–18.

Plakidou-Dymock, S., P. J. Warner and I. J. Higgins. 1991. "Plasmid of Alkane-Utilising *Arthrobacter*," *FEMS Microbiol. Lett.*, 80:51–56.

Saito, H. and K. Miura. 1963. "Preparation of Transforming Deoxyribonucleic Acid by Phenol Treatment," *Biochem. Biophysic. Act.*, 72:619–629.

Shingler, V., J. Powlowski and U. Marklund. 1992. "Nucleotide Sequence and Functional Analysis of the Complete Phenol/3,4-Dimethylphenol Catabolic Pathway of *Pseudomonas* sp. Strain CF600," *J. Bacteriol.*, 174:711–724.

Stotzky, G. and H. Babich. 1986. "Survival and Genetic Transfer by Genetically Engineered Bacteria in Natural Environments," *Advances in Appl. Microbiol.*, 31:93–138.

Tait, R. C., T. J. Chose, R. C. Lundquist, M. Hagiya, L. Rodriguez and C. I. Kado. 1983. "Construction and Characterization of a Versatile Broad Host Range DNA Cloning System for Gram-Negative Bacteria," *Biotechnol.*, 1:269–275.

Vieira, J. and J. Messing. 1982. "The pUC-Plasmid, an M13mp7-Derived System for Insertion Mutagenesis and Sequencing with Synthetic Universal Primers," *Gene*, 19:259–268.

CHAPTER 5

Breeding of Genetically Engineered Microorganisms for Wastewater Treatment—A Few Case Studies

Genetic manipulation of degrading bacteria may offer several advantages for the removal of xenobiotic or recalcitrant compounds, a major issue in wastewater treatment. McClure et al. (1991) suggested the following:

- The hybrid catabolic pathway of xenobiotic compounds may cause the breakdown of novel substrates.
- Manipulation of regulatory genes can raise the constitutive expression of degrading genes.
- The host range can be extended to propagate a specific catabolic gene in a mixed community of various bacteria.
- Amplification of a rate-determining step of the catabolic pathway may enhance degradation rates.
- Prevention of the formation of toxic intermediates by the genetic block may have desirable effects on degradation activity.

Many studies have demonstrated these advantages. Reports on the studies focused on the gene-mediated degradation and its biochemical and genetic aspects, but none evaluated the GEMs as the functional microorganisms. This chapter will describe a few case studies of the breeding of GEMs applicable to wastewater treatment, using genes which degrade aromatic compounds.

The first study describes the manipulation of the expression of the C230-encoding gene, a conventional way to evaluate GEMs. The second and third studies evaluated practical applications of constructed GEMs for which the catabolic range was extended, and for which a rate-determining step was amplified.

5.1 INCREASE OF GENE EXPRESSION

5.1.1 Host/Vector Systems

To evaluate the increase, by genetic manipulation, in the expression of genes responsible for degrading xenobiotic compounds, a C230-encoding gene, *pheB*, which was cloned from the chromosomal DNA of a phenol-degrading bacterium, *P. putida* BH (see Section 4.3.), was used as a model system. Since 2-HMS, which is produced from catechol by the catalysis of C230, is a strongly absorbing substance at 375 nm (1 μmol/mL[144 mg/L] of 2-HMS in phosphate buffer gives an OD_{375} of 44.1 in a 10-mm glass cuvette), the C230 activity can be assayed easily and precisely by spectrophotometry. Consequently, this system is considered suitable for the evaluation of gene expression in various combinations of hosts and vectors.

Bacterial strains and plasmids used in this comparative study of gene expression are listed in Table 5.1. Cloning vectors pUC19 and pKT230 were

TABLE 5.1. Bacterial strains and plasmids.

Strain or plasmid	Phenotype/Genotype[a]	Source or reference[b]
E. coli		
JM103	thi$^-$, *hsdR*	
C600	thr$^-$, leu$^-$, thi$^-$, *hsdR hsdM*	
P. putida		
BH	Phe$^+$, Ben$^+$	Isolated from activated sludge
BH–1	Phe$^-$, Ben$^+$	Derived from BH
KT2440	Ben$^+$, *hsdR*	Bagdasarian et al., 1981
PpG1064	Phe$^+$, Ben$^+$, trp$^-$	Dunn and Gunsalus, 1973
Plasmid		
pUC19	*Ap*, lac/p, high copy number	Vieira and Messing, 1982
pKT230	*Sm*, *Km*, broad host range	Bagdasarian et al., 1981
pBH100	*Ap*, catechol 2,3-oxygenase	Cloning of *pheB* from the chromosome of BH in pUC19. Fujita et al., 1991
pBH500	*Sm*, catechol 2,3-oxygenase	*Hin*dIII–*Bam*HI fragment of pBH100 (5.65kb) containing *pheB* in pKT230.

[a]Requirement for thi: thiamine, thr: threonine, leu: leucine, and trp: tryptophan; Capability of growing on Phe: phenol, Ben: benzoate, Nah: napthalene, and Sal: salicylate; *hsdR*: host-specific restriction, *hsdM*: host-specific modification, lac/p: *lac* promotor.
[b]References are shown in Chapter 4 and Chapter 5.

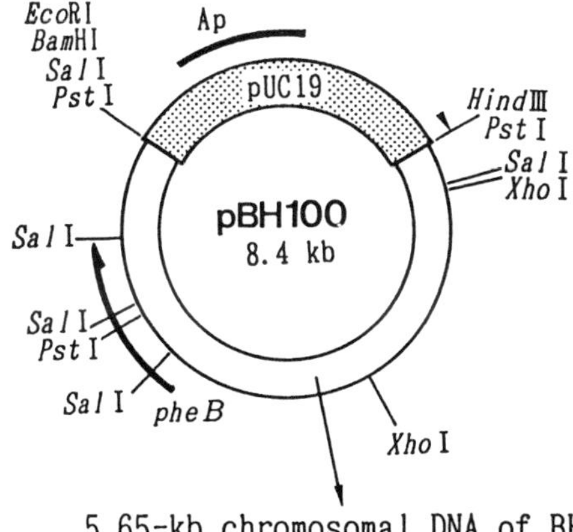

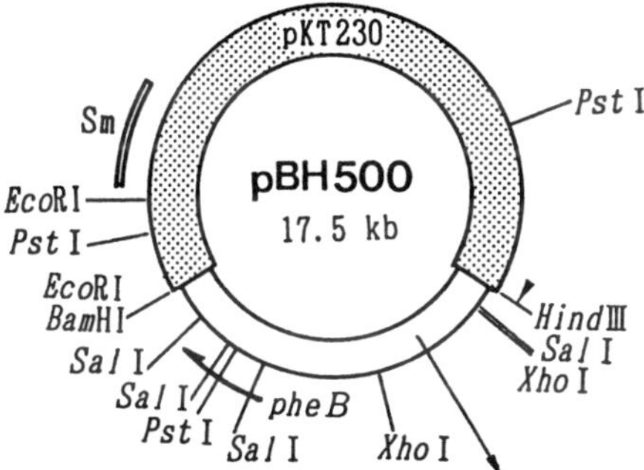

FIGURE 5.1. Structure of pBH100 and pBH500.

used; their structures are shown in Figure 4.6 and Figure 4.8, respectively. The recombinant plasmid pBH100 was constructed by cloning the 5.65-kb DNA fragment (4.25-kb plus 1.40-kb *Pst*I fragment) containing the C230 encoding *pheB* gene of *P. putida* BH into the *Pst*I site of pUC19 (Fujita et al., 1991). The pBH500 was constructed by the replacement of the small *Hin*dIII-*Bam*HI fragment with a 5.65-kb *Hin*dIII-*Bam*HI fragment of pBH100 DNA containing *pheB*. The structures of these two recombinant plasmids are diagramed in Figure 5.1.

The pBH100, constructed by a narrow host range vector for *E. coli* pUC19, was introduced into *E. coli* hosts (C600 and JM103) by transformation. The pBH500, constructed by a broad host range vector, was introduced into *E. coli* C600 and *P. putida* strains by transformation or triparental conjugation with a mobilizer *E. coli* C600(RP4). *P. putida* strains used as hosts were PpG1064, BH-1, and KT2440. *P. putida* PpG1064(NAH), a wild strain isolated by Dunn and Gunsalus (1973), was treated simultaneously with sodium dodecyl sulfate (0.05%) and mitomycin C (1.5 μg/mL) to obtain a plasmid-free derivative *P. putida* PpG1064. *P. putida* BH-1 is a mutant deficient in the ability to use phenol. It is derived from *P. putida* BH through mutagenesis with irradiation by ultraviolet light.

5.1.2 Expression of C230 Gene

The comparison of the expression of the *pheB* gene was carried out by measuring C230 activity of the crude cell extract of GEMs as follows.

The GEMs were grown with LB broth containing peptone 10 g, yeast extract 5 g, and NaCl 5 g in 1 L of water on a rotary shaker (160 rpm, 30°C), and harvested by centrifugation at 4°C. The cell pellets were washed twice in cold 50 mM phosphate buffer, then disrupted sonically at 0°C in the suspension with the buffer. Cellular debris was removed by centrifugation at 25,000 × *g* for 20 min at 4°C, and the clear supernatant was used immediately for enzyme assay after adding acetone at a final concentration of 10% for the stabilization of C230. Catechol was added to the cell-free crude extract, and C230 activity was assayed by measuring 2-HMS formation (Nakazawa and Yokota, 1973). The C230 activity was expressed as μmol of 2-HMS produced/min/mg of protein contained in the crude extract (units/mg-protein). To evaluate the maximum C230 activity of the GEMs, cells were sampled periodically for the enzyme assay.

TABLE 5.2. C230 activities of GEMs.

Plasmid	Host strain	Growth medium[a]	C230 activity[b]
none	*P. putida* BH	LB + Phenol	100.4
	P. putida BH	LB + Benzoate	8.9
pBH100	*E. coli* JM103	LB	233.7
	E. coli C600	LB	854.6
pBH500	*E. coli* C600	LB	73.0
	P. putida PpG1064	LB	60.8
	P. putida BH-1	LB	25.1
	P. putida KT2440	LB	23.9

[a]Both phenol and benzoate are added at concentration of 500 mg/L.
[b]Units/mg-protein.
Reprinted from *Water Research, Vol. 25,* Fujita, Ike and Hashimoto, "Feasibility of Wastewater Treatment Using Genetically Engineered Microorganisms," p. 982, 1991, with kind permission from Pergamon Press Ltd., Headington Hill Hall, Oxford OX3 0BW, UK.

Table 5.2 shows the C230 activity of genetically engineered *E. coli* and *P. putida* harboring pBH100 or pBH500. The C230 level in the parental wild strain *P. putida* BH is also contained in the table.

The wild strain *P. putida* BH expressed the *pheB* gene and showed the C230 activity at 100 units/mg-protein only when induced by phenol. When cultivated with benzoate, it showed C230 activity at a much lower level (9 units/mg-protein), indicating that phenol is a specific substrate to induce C230. On the other hand, GEMs carrying the recombinant plasmid pBH100 or pBH500 expressed the *pheB* gene consistently without any inducers, such as phenol. It seems likely, therefore, that expression of the *pheB* gene was governed by constitutive transcription from the promotors on the vectors used for the GEM construction. The pUC19 and pKT230, respectively, carry a single transcriptional promotor, *lac* promotor and *Km* promotor. It has been demonstrated that the cloned DNA fragment containing *pheB* contains no putative promotor regions for *Pseudomonas* (data not shown).

Table 5.2 shows that the C230 activity of the GEMs depended upon the host/vector systems. The C230 activity of *E. coli* C600 (pBH100) was 8.5 times higher than that of the phenol-induced wild strain. *E. coli* JM103 carrying pBH100 also showed 2.3 times higher activity of the C230. This suggested that gene amplification by the high copy number recombinant plasmid pBH100 constructed with pUC19 caused the enhancement of the C230 activity. On the other hand, any genetically modified *E. coli* and *P. putida* strains harboring the recombinant plasmid pBH500 could not show higher expression of the *pheB* gene than *P. putida* BH. Even the maximum C230 activity of the pBH500-harboring GEM was approximately 70% of the wild strain. This lower C230 activity of the GEMs may depend on the lower copy number of the plasmid pBH500 constructed with pKT230 and pBH100 constructed with pUC19. From that it is generally considered that the pUC19 plasmid is present in 50–60 copies per cell in exponentially growing cells. The pKT230 plasmid is thought to number 15–20 copies per cell. The copy number of the recombinant plasmid carrying an inserted foreign DNA fragment is a little lower than that of the corresponding vector plasmid. The copy number of the *pheB* gene in the GEMs harboring pBH100 was considered to be approximately 3–4 times higher than that in the GEMs harboring pBH500.

At any rate, it seems we can increase or enhance enzyme activity for xenobiotic degradation by using a high copy number plasmid vector to construct the GEMs. Thus, it was also suggested that the application of such GEMs can lead to the efficient removal of xenobiotic compounds in wastewater treatment processes.

The expression of the *pheB* gene was considerably affected by the host strain. The C230 activity expressed by *pheB* carried on pBH100 in *E. coli*

C600 was 855 units/mg-protein, while that in *E. coli* JM103 was 234 units/mg-protein (approximately 30%). Among the pBH500-harboring GEMs, the highest C230 activity (73 units/mg-protein) was shown in *E. coli* C600, and that in *P. putida* strains was 24–60 units/mg-protein. Although the reasons for these differences in the C230 activity were not made clear, one may be due to the regulatory systems for the expression of the *pheB* gene. For example, the C230 activity coded on pBH500 was higher in an *E. coli* host than in *P. putida* hosts in spite of the fact that the *pheB* gene was cloned from the chromosome of *P. putida* BH. This suggests that the *Km* promotor of pKT230, derived from the DNA of *E. coli,* may be transcribed more efficiently under the genetic background of *E. coli* than under that of *P. putida.* If this theory is right, a vector which has a *Pseudomonas*-derived strong promotor may enhance the C230 activity in *P. putida* host strains, even if it cannot be maintained in as high a copy number as in the host cell.

There appeared to be no special merits to the use of pseudomonads (aerobic gram-negative rod-shaped bacteria) as recipients of the catabolic gene. However, from the practical viewpoint, pseudomonad hosts are considered suitable for wastewater treatment because they seem to be dominant genera in the treatment process (see Chapter 2). So, it is necessary to choose the optimal combination of pseudomonads and broad host range vectors for genetic manipulation in order to create a GEM with higher degradation activity, and one that may be dominant in the actual processes.

5.1.3 Conclusions and Considerations

The manipulation of the expression of cloned C230-encoding gene *pheB* was investigated and evaluated. The advantages of manipulation of the xenobiotic degrading gene observed in this study may be summarized as follows:

- Although the wild strain *P. putida* BH expressed the *pheB* gene only when cultivated with phenol, the GEMs showed the C230 activity consistently. This may depend on the constitutive transcription of the gene from the promotor present on the vector pUC19 or pKT230.
- The C230 activity was enhanced to approximately 8.5 times the level of phenol-induced *P. putida* BH in *E. coli* C600 with the pUC19 vector. This may be due to the high number of gene copies in the host cell.
- The *pheB* gene was introduced into a variety of bacterial strains with the broad host range vector pKT230. All strains expressed the

gene. The pBH500, which was constructed with the RSF1010-derived vector, pKT230, may be introduced into a much wider range of hosts than that used here.

These results suggest the high potential for manipulating degradative genes to enhance the removal efficiency of xenobiotics in wastewater treatment by using GEMs. However, the data showed a few problems in GEM breeding. For example, the GEMs harboring recombinant plasmid pBH500 showed lower C230 activity than the wild strain. It was also observed that the expression of *pheB* was lower in *P. putida* strains than in an *E. coli* strain. Since gene expression ordinarily depends on the activity of a promotor, and on the number of gene copies per cell, the screening of a high copy number vector containing the most efficient promotor transcribed in the given hosts, or of a host strain suitable for the given high copy number expression vector, is important in the creation of useful GEMs. However, please note that in this study, the availability of GEMs was evaluated only by a specific gene expression in a cell-free system. This technique may not directly lead to the evaluation of GEMs in the actual wastewater treatment applications.

5.2 EXTENSION OF CATABOLIC RANGE

5.2.1 Catabolic Pathway of Aromatic Compounds

For breeding GEMs capable of degrading xenobiotic compounds, it is essential to elucidate their catabolic pathways and the participating enzymes. Many studies have dealt with such catabolic pathways and enzymes from biochemical aspects, revealing some interesting features of microbial degradation. Among these commonly observed findings, a most interesting one is that many degradative pathways for various complex organic compounds converge into limited common intermediates. Moreover, the distinct pathways often employ isofunctional catabolic enzymes coded on different catabolic genes, and such isofunctional enzymes exhibit varying substrate specificities.

Figure 5.2 shows several pathways for the bacterial degradation of a variety of aromatic compounds. As mentioned above, it also shows the convergence of catabolic pathways. In the biodegradation of aromatic compounds, five phases can be distinguished:

(1) entry into the cell, often by free diffusion but sometimes by the specific transport mechanisms which exist for even such simple compounds as benzoate and mandelate

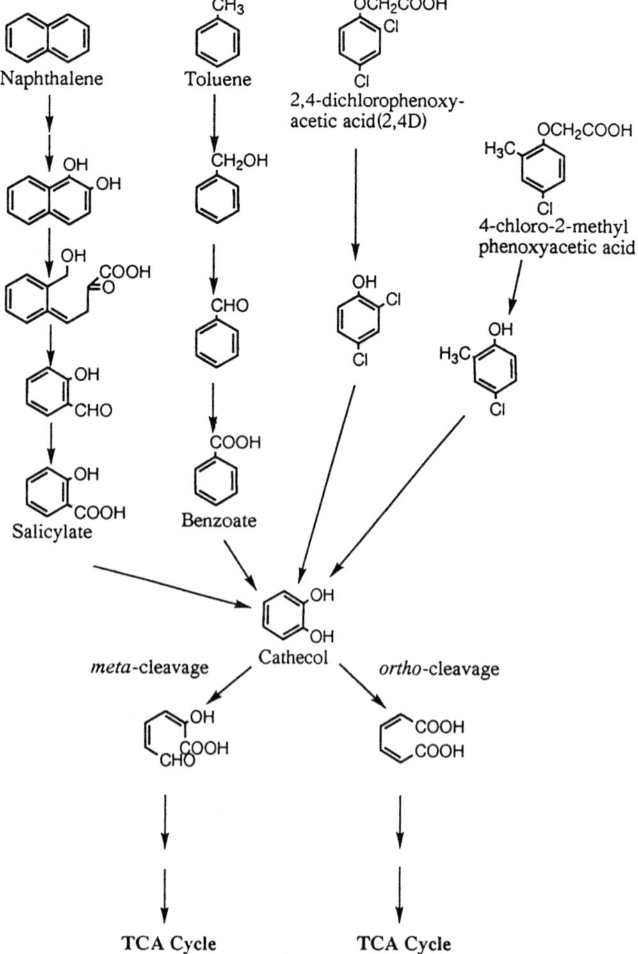

FIGURE 5.2. Convergence of catabolic pathway for aromatic compounds.

(2) manipulation of the side chains and formation of substrates for benzene ring cleavage
(3) benzene-ring cleavage
(4) conversion of the products of ring cleavage into simpler intermediates which may appear in the TCA cycle
(5) utilization of such simple compounds (Fewson, 1981)

Catechol (or catechol substitutes) or a few other di- or trihydroxy phenols (portocatechuic acid, homoprotocatechuic acid, gentistic acid, gallic acid, etc.) are central intermediates into which many substrates are transformed. Key intermediates, such as catechols or protocatechuic acid, are also common substrates for ring cleavage. In general, modification of substituents occurs before ring cleavage, but this is not always the case. In the case of *ortho*-dihydroxy phenols, there can be intradiol or extradiol cleavage, which are called *ortho*- or *meta*-cleavage pathways, respectively. The enzymes responsible for the ring fission, especially for *meta*-cleavage, exhibit broad substrate specificities, therefore allowing degradation of a wide range of compounds. For example, C230, which catalyzes the *meta*-fission of catechol, coded on the *xylE* gene, can transform 3-methylcatechol, 4-methylcatechol, 3-chlorocatechol, 4-chlorocatechol, protocatechuic acid, protocatechualdehyde, and pyrogallol (Nozaki, 1970). Although another isofunctional C230 specified by *pheB* could degrade all such compounds, the specific activities relative to catechol itself were different (Table 5.3). On the other hand, catechol 1,2-oxygenase, which catalyzes the *ortho*-fission of catechols from different microorganisms, shows a fairly strict, but generally not an absolute, specificity for their substrates. So the substrate specificity of the enzymes may depend upon the bacterial strains and their catabolic pathways. As a whole, the enzymes concerning upper reactions in a certain catabolic pathway for aromatic compounds (e.g., naphthalene to salicylate, 2,4-dichlorophenoxyacetic acid to 2,4-dichlorophenol, etc.) tend to have narrower specificities than those which catalyze lower reactions in the pathway.

5.2.2 Genetic Manipulation for Extension of Catabolic Range

Although various bacteria in soil or water environments can degrade and grow upon a wide variety of aromatic compounds through the above-

TABLE 5.3. Substrate specificity of C230.

	Relative activity[a]	
Substrate	*pheB*	*xylE*[b]
Catechol	100.0	100.0
3-Methylcatechol	60.8	30.6
4-Methylcatechol	299.0	89.5
4-Chlorocatechol	274.0	93.1
Protocatechuate	0.20	0.26
Protocatechualdehyde	1.57	1.53
Pyrogallol	0.10	0.33
Phlorogulucinol	0.0	0.0

[a]Percent activity toward catechol.
[b]Referring to Nozaki (Nozaki, 1970).

mentioned catabolic pathways, each individual strain usually can utilize only a few compounds simultaneously and/or efficiently. This is a particularly important problem for the bioremediation of polluted environments or wastewater treatments, since a variety of xenobiotic compounds, including aromatic compounds, may be present or discharged in a complex mixture in such situations. Difficulty in the simultaneous degradation of not only aromatic compounds but also other recalcitrant compounds by certain degrading microorganisms involves the following genetic or biochemical problems of catabolic pathways:

- Individual degrading microorganisms possess only a few distinct pathways for degradation of xenobiotic or recalcitrant compounds. This may be because the total size of the bacterial genome (genetic information) is not large enough to accommodate many extra functions, a condition which may not confer an advantage under ordinary environmental conditions.
- Even when many types of catabolic pathways coexist in a microorganism, only one (or a few) type(s) is active under any given condition. This suggests complex regulation systems for the expression of catabolic genes or pathways. It may be possible that expression of one catabolic pathway or gene prevents or lowers that of another coexisting in a particular microorganism.
- During degradation of a certain compound, dead-end products or reactive intermediates are formed which have inactivating or lethal effects upon the corresponding degrading microorganisms. Accordingly the degrading activities stop.

Such problems may be solved by genetic manipulation of catabolic pathways. For example, the lack of an essential catabolic enzyme for the target pollutant in a microorganism can be supplemented by introducing the corresponding catabolic genes cloned from other microorganisms to establish a novel catabolic pathway. The formation of a dead-end product may also be prevented by genetic block of the pathway. Therefore, the combination of these genetic manipulations can generate a superior microorganism capable of degrading a wide variety of xenobiotic compounds simultaneously and efficiently.

A most sophisticated example of the construction of a GEM responsible for degradation of aromatic compounds by extension of the catabolic range was reported by Rojo et al. (1987). Their report described the construction of a GEM able to degrade and grow on mixtures of chloro- and methylaromatics that were toxic even for bacteria capable of degrading the individual components of the mixture. The strategy of this work was the "patchwork assembly of new pathways by the judicious combination of enzymes recruited from different pathways and different organisms."

The original or parental bacterial strain selected for the genetic manipulation was *Pseudomonas* sp. B13, which had a narrow spectrum of chloro- and methylaromatics. This strain has a well-characterized *ortho*-cleavage route for the degradation of 3-chlorobenzoate (3CB) and 4-chlorophenol (4CP) and originally lacked detectable *meta*-cleavage activity.

First, they recruited the TOL plasmid pWWO encoded enzymes – toluate dioxygenase and dihydroxycyclohexadiene carboxylate dehydrogenase – which catalyze the conversion of methylbenzoates to methylcatechols (Harayama et al., 1986). The cistrons *xylXYZ* and *xylL* of pWWO – coding for toluate dioxygenase and dihydroxycyclohexadiene carboxylate dehydrogenase, respectively – were cloned and inserted into transposon *Tn5*, and the resulting hybrid transposon (recombinant transposon) was then transported into the chromosome of *Pseudomonas* B13. In addition to the above four cistrons (*xylXYZ* and *xylL*), whose promotor *Pm* and cistron *xylS* of pWWO specifying the regulatory protein which positively controls *Pm* (by accelerating the transcription of *xylXYZ* and *xylL*) were contained in the hybrid transposon. One derivative of *Pseudomonas* B13 carrying the hybrid transposon was selected and designated *Pseudomonas* sp. FR1. This derivative could completely degrade 4-chlorobenzoate (4CB) in addition to 3CB and 4CP due to relaxed specificities of pWWO-derived catabolic enzymes; 4CB was transformed into 4-chlorocatechol by the catalysis of such TOL enzymes, then completely degraded through the *ortho*-route coded on the chromosome of *Pseudomonas* B13 (FR1). *Pseudomonas* FR1 was able not only to utilize 4CB but also to transform 3-methylbenzoate and 4-methylbenzoate (3MB and 4MB) to 4-carboxymethyl-2-methylbut-2-ene-1,4-olide (2-methyl-2-enelactone) and 4-carboxymethyl-4-methylbut-2-ene-1,4-olide (4-methyl-2-enelactone) respectively by co-metabolism. However, *Pseudomonas* B13(FR1) could grow on 4-carboxymethyl-3-methylbut-2-ene-1,4-olide (3-methyl-2-enelactone), but not on the 2-methyl-2-enelactone and 4-methyl-2-enelactone as a sole carbon and energy source.

Next, *Pseudomonas* FR1 was supplemented with the enzymes that transform methyl-2-enelactones into TCA cycle intermediates so as to establish an *ortho* pathway for complete degradation of methylbenzoates. In this step, the donor of genes responsible for methylbenzoate degradation through *ortho* pathway was *Alcaligenes eutrophus* JMP134 (Boyer and Roulland-Dussoix, 1969). This strain has a pathway for the degradation of 4-methylcatechol, in which 4-methyl-2-enelactone formed from 4-methylcatechol is isomerized into 3-methyl-2-enelactone, then transformed into 4-methyl-3-oxoadipic acid, and further to TCA cycle intermediates. The total genomic library of *A. eutrophus* JMP134 was constructed with the cosmid vector pLAFR3 in *E. coli* HB101 to clone the corresponding genes. The bank of hybrid cosmids was mass transferred by triparental conjugation into *Pseudomonas* FR1, and transconjugants which could grow on 4MB as a sole source of carbon and energy were selected. One of the transconjugants was studied further, and the hybrid cosmid, containing a 26-kb DNA insertion from the chromosome of *A. eutropus* JMP134, was designated pFRC20P. The analysis of pFRC20P showed that the cloned DNA fragment coded for the enzyme catalyzes the conversion of 4-methyl-2-enelactone into 3-methyl-2-enelactone. Consequently, the constructed transconjugant *Pseudomonas* FR1 (pFRC20P) could completely degrade 4MB. In the study, 4MB was transformed into 4-methyl-2-enelactone in *Pseudomonas* FR1, and degraded completely via 3-methyl-2-enelactone, which was produced by the enzyme from *A. eutrophus* JMP134.

Finally, mutational activation of phenol hydroxylase of *Pseudomonas* B13 was employed to extend the catabolic range of *Pseudomonas* FR1 (pFRC20P) to chloro- and methylphenols. Theoretically, *Pseudomonas* FR1 (pFRC20P) could grow on 4-methylphenol (4MP). *Pseudomonas* B13 has the phenol hydroxylase, which could transform not only phenol into catechol, but also 4MP into 4-methylcatechol in cell-free extract, and *Pseudomonas* FR1 (pFRC20P) mineralized 4MB via 4-methylcatechol. The recombinant should possess all the enzymes necessary for the mineralization of 4MP. After adaptation of this GEM, spontaneous mutants, *Pseudomonas* FR1 (pFRC20P)-1 and *Pseudomonas* FR1 (pFRC20P)-2, which could grow on 4MP were selected. Since the production of the phenol hydroxylase of *Pseudomonas* B13 was inducible by phenol, the lack of the ability to degrade 4MP by *Pseudomonas* FR1 (pFRC20P) was believed to arise from the fact that the phenol hydroxylase was not induced by 4MP. Therefore, the mutation which occurred in *Pseudomonas* FR1 (pFRC20P)-1 and *Pseudomonas* FR1 (pFRC209P)-2 might be associated with the induction system of the phenol hydroxylase. Measurement of the phenol hydroxylase levels in *Pseudomonas* FR1 (pFRC20P) and its mutant bacterial strains grown either in succinate, phenol, or 4MP showed that little or no activities were observed in succinate- or 4MP-grown cells of the parental strain, and in succinate-grown cells of the mutants, and that high levels of activity were expressed in phenol-grown cells of both parental and mutant strains, and in 4MP-grown cells of the mutants.

In conclusion, a narrow catabolic range of chloro- and methylaromatics in *Pseudomonas* B13 was extended to an extremely broad range by various genetic manipulations of catabolic pathways. The ability to degrade 4MB, 4CP, and 4MP were added in addition to those of 3CP and 4CP. From this result, Rojo et al., emphasized that the evolutionary process of the catabolic pathways for novel xenobiotic compounds can be considerably accelerated by rational genetic manipulation. Although the evolution of entire catabolic pathways may proceed very slowly under natural environmental conditions, experimental evolution of pathways may offer a major advantage in that laboratory selection conditions can be custom designed for each of the individual changes required.

The following section describes an example of the construction of an aromatic compound degrading GEM in which the catabolic range was extended. The study describes a simpler procedure in which the new catabolic enzyme or pathway was conferred on the parent bacterium. The constructed GEM also showed higher degradation rates than the wild strain in simultaneous degradation of aromatics, maybe via a coordinated regulation of pathways.

5.2.3 Simultaneous Degradation of Phenol and Salicylate

Genetic manipulation was performed upon a phenol and benzoate degrading bacterium, *P. putida* PpG1064. Originally, *P. putida* PpG1064 harbored the naturally occurring catabolic plasmid NAH (Dunn and Gunsalus, 1973), which specified naphthalene or salicylate degradation [the wild strain *P. putida* PpG1064 (NAH)]. The catabolic pathway of salicylate coded on NAH plasmid was altered by genetic manipulation in this study, and the degradation property of the resultant GEM for aromatic compounds (phenol and salicylate) was compared with that of the wild strain.

The catabolic pathways of the wild strain, *P. putida* PpG1064 (NAH), for phenol and salicylate are shown in Figure 5.3. Phenol is metabolized through the *ortho*-cleavage pathway for catechol, of which the corresponding enzymes are coded on the chromosome of the host strain, *P. putida* PpG1064. Salicylate, on the other hand, is first transformed into catechol by the catalysis of salicylate hydroxylase, specified by the *nahG* gene, then degraded through the *meta*-cleavage pathway for catechol. All the enzymes associated with such salicylate degradation are coded on the catabolic plasmid, NAH. Therefore, *P. putida* PpG1064 (NAH) has both *ortho*- and *meta*-cleavage pathways for catechol degradation produced from phenol and salicylate, respectively. But the recovered derivative, *P. putida* PpG1064, has an *ortho*-cleavage pathway only for catechol degradation.

We designed the GEM for degrading both phenol and salicylate through the only *ortho*-cleavage pathway coded on the chromosome of *P. putida* PpG1064. The *nahG* gene was isolated from the NAH plasmid DNA and introduced into cured strain of *P. putida* PpG1064 through the following genetic manipulation procedures. It was known that the *nahG* gene was contained in the 3.1-kb *Hin*dIII fragment of the NAH plasmid DNA—*Hin*dIII generates 16 DNA fragments from the NAH plasmid (Lehrbach et al., 1984)—this fragment was inserted into the corresponding site of the cloning vector, pBR325, and the resultant recombinant plasmid pHF100 was

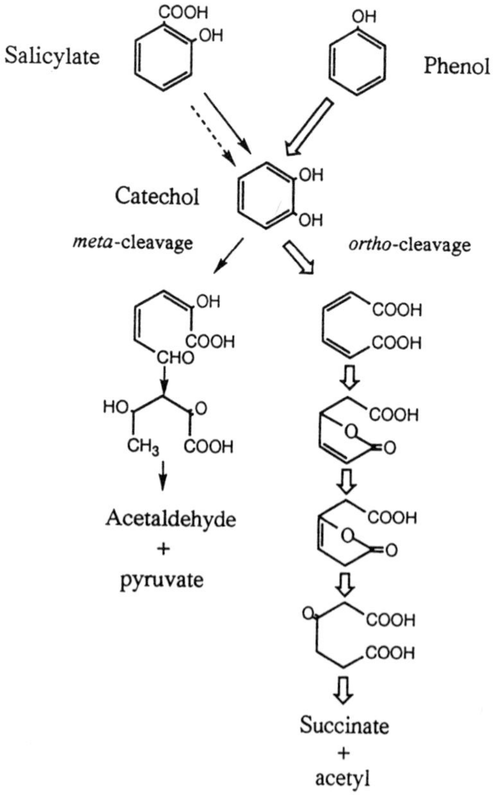
Salicylate
COOH
OH
OH
Phenol
Catechol
OH
OH
meta-cleavage
ortho-cleavage
OH
COOH
CHO
COOH
COOH
HO
O
CH3
COOH
Acetaldehyde
+
pyruvate
COOH
O
O
COOH
O
O
O
COOH
COOH
Succinate
+
acetyl

FIGURE 5.3. Phenol and salicylate catabolic pathway in *P. putida* PpG1064 (NAH) and *P. putida* PpG1064 (pHF400). ⇨ coded on the chromosome of PpG1064; → coded on NAH-plasmid; --→ coded on pHF400.

transformed into *E. coli* C600. The transformants could convert salicylate into catechol when cultivated in LB broth supplemented with salicylate. Their presence was determined by their ability to turn brown on the medium due to the accumulation of catechol, which was spontaneously oxidized into a brownish compound. However, the transformants were not able to grow on salicylate as a sole carbon and energy source, because of the lack of a catechol catabolic pathway in the host strain *E. coli* C600.

Since the pHF100 constructed with the narrow host range plasmid vector pBR325 could not transfer into *P. putida* PpG1064, the 3.1-kb *Hin*dIII DNA fragment cut from pHF100 was ligated with the *Hin*dIII-digested secondary vector pKT230, which has a broad host range and can be introduced into a wide variety of gram-negative bacteria, including *P. putida* strains. The resultant recombinant plasmid, designated pHF400, was first transformed into *E. coli* C600 (transformants were selected in the same manner as for pHF100), then transferred into *P. putida* PpG1064 by triparental conjugation with a helper *E. coli* C600 (RP4). The transconjugants were selected on a medium containing salicylate as a sole carbon source, and designated *P. putida* PpG1064 (pHF400). Thus, the ability to completely degrade salicylate was observed in *P. putida* PpG1064, which was originally able to utilize phenol and benzoate. The metabolic pathways of *P. putida* PpG1064 (pHF400) for phenol and salicylate are also shown in Figure 5.3.

The growth and degradation of phenol and salicylate by the GEM, *P. putida* PpG1064 (pHF400), and the wild strain, *P. putida* PpG1064 (NAH), were investigated and compared with each other by batch cultures using the minimal salts medium. One milliliter of overnight culture of each strain in LB broth was inoculated into 100 mL of minimal salts media containing phenol at 500 mg/L (phenol medium), salicylate at 400 mg/L (salicylate medium), or both phenol and salicylate at 300 mg/L respectively (phenol-salicylate medium) in a 300-mL Erlenmeyer flask. Flasks were rotated on a rotary shaker at 160 rpm at 30°C for phenol and salicylate degradation tests.

Since genetic manipulation did not cause any change in the phenol catabolic pathway by *ortho*-cleavage of *P. putida* PpG1064, the phenol degradation property of the GEM was considered the same as that of the host strain, *P. putida* PpG1064. However, the results of the degradation tests of *P. putida* PpG1064 (pHF400) and *P. putida* PpG1064 exhibited a small difference in the growth and phenol degradation in phenol medium (Figure 5.4); the GEM showed a little slower growth and phenol degradation than the host strain.

This might be due to the burden of plasmid-possession by the GEM. An excess production of protein coded on pHF400, which was not essential for phenol degradation, seemed to load the GEM. Moreover, although phenol degradation by the wild strain, *P. putida* PpG1064 (NAH), was monitored, no distinct difference between the wild strain and the GEM, *P. putida* PpG1064 (pHF400), was recognized. This suggested that the burden of the NAH plasmid-loaded *P. putida* PpG1064 as well as in the case of the GEM (where the pHF400 loaded the host)—or that the *meta*-cleavage pathway coded on the NAH plasmid—had a deleterious effect on the catechol degradation produced from phenol.

Cell growth and salicylate degradation of the GEM, *P. putida* PpG1064 (pHF400), and the wild strain, *P. putida* PpG1064 (NAH), in a salicylate medium were compared in Figure 5.5. It is apparent from the figure that the patterns of growth and salicylate degradation of both strains were similar to each other, but that the wild strain grew and degraded salicylate a little faster than the GEM did.

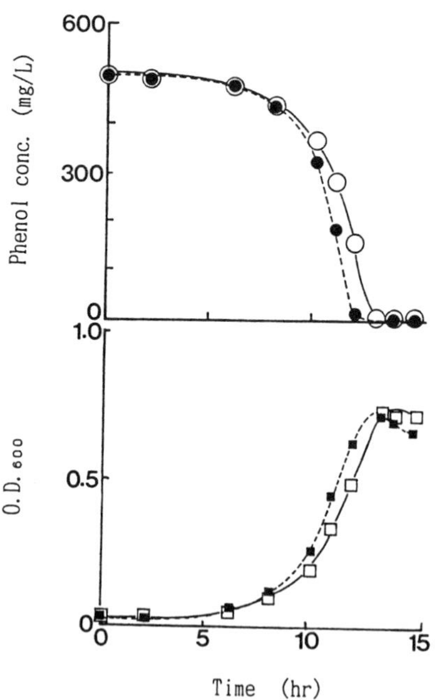

FIGURE 5.4. Phenol degradation by *P. putida* PpG1064 and *P. putida* PpG1064 (pHF400). ●, phenol removal by and ■, cell growth of PpG1064; ○, phenol removal by and □, cell growth of PpG1064 (pHF400).

The wild strain might degrade salicylate by a well-coordinated expression of catabolic enzymes for the *meta*-cleavage pathway which are coded on the naturally occurring NAH plasmid, and which are all transcribed as a single positively regulated operon induced with salicylate. The GEM seemed to transform salicylate into catechol, first by the expression of the *nahG* gene coded on the recombinant plasmid, pHF400, which has only one constitutively transcriptional promotor (for *Km*) on it, and then completely degraded catechol by the *ortho*-cleavage pathway coded on the chromosome of the host, *P. putida* PpG1064. The reason for the slightly slower growth and salicylate degradation of the GEM compared to the wild strain might depend upon a double-phased and probably ill-coordinated catabolic pathway for salicylate in the GEM. (The first phase is the conversion of salicylate into catechol, and the second phase is further degradation of catechol into TCA cycle intermediates.) In short, it was concluded that the coordinated single regulatory gene expression system is preferable to the combination of two distinct expression systems for salicylate degradation.

Figure 5.6 shows the course of cell growth and substrate concentrations in degradation tests using phenol-salicylate medium. The results for the GEM, *P. putida* PpG1064 (pHF400), and for the wild strain, *P. putida* PpG1064 (NAH), are compared in the figure. As shown in the figure, although both strains completely degraded phenol and salicylate simultaneously and effectively, the patterns of degradation of substrates were different. The wild strain degraded salicylate faster than phenol; the reverse was true in the GEM.

The reason for this difference between the GEM and the wild strain in the simultaneous degradation of phenol and salicylate is thought to lie in the regulatory mechanisms for the expression of corresponding catabolic genes. In the wild strain, *P. putida* PpG1064 (NAH), the expression of the *nahG* gene and the following genes responsible for *meta*-cleavage pathway for catechol coded on the NAH plasmid were highly induced by salicylate from the start of the test, while the expression of the *nahG* gene in the GEM, *P. putida* PpG1064 (pHF400), was constitutive. However, the catechol catabolic enzymes from the *ortho*-cleavage pathway might not be induced directly by salicylate. Thus, the degradation of salicylate by the

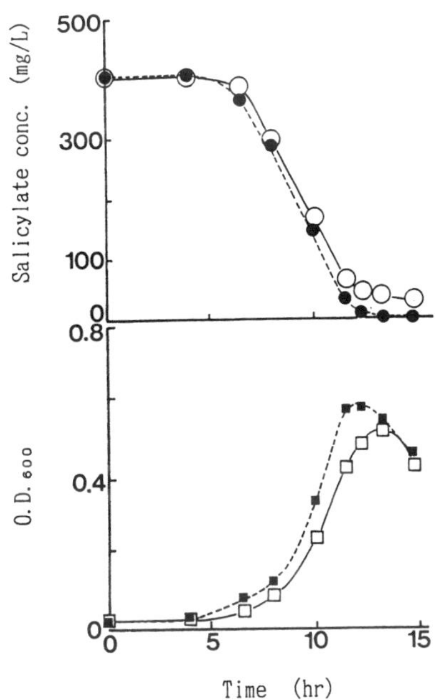

FIGURE 5.5. Salicylate degradation by *P. putida* PpG1064 (NAH) and *P. putida* PpG1064 (pHG400). ●, salicylate removal by and ■, cell growth of PpG1064 (NAH); ○, salicylate removal by and □, cell growth of PpG1064 (pHF400).

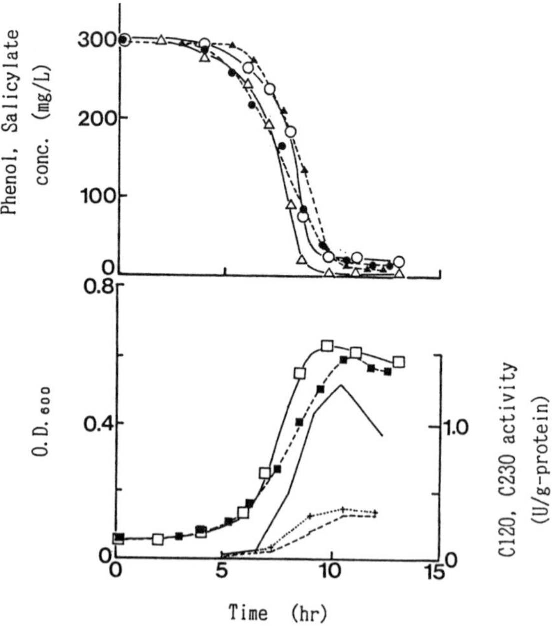

FIGURE 5.6. Simultaneous degradation of phenol and salicylate by *P. putida* PpG1064 (NAH) and *P. putida* PpG1064 (pHF400). ●, salicylate; ▲, phenol removal by and ■, cell growth; +, C120; - - -, C230 activity of PpG1064 (NAH); ○, salicylate; △, phenol removal by and □, cell growth; —, C120 activity of PpG1064 (pHF400).

GEM occurred a little later than that by the wild strain. On the other hand, the *ortho*-cleavage pathway for catechol is generally considered to be induced not by the primary substrates (phenol, benzoate, salicylate, etc.) but by the intermediates converted from catechol (cis,cis-muconate, β-ketoadipate, etc.). Therefore, the *ortho*-cleavage pathway coded on the chromosome of *P. putida* PpG1064, which might be used for phenol degradation, was induced at a higher level in the GEM than in the wild strain, probably because its inducers such as cis,cis-muconate and β-ketoadipate, might be produced and accumulated more in the GEM, which has only the *ortho*-cleavage pathway for degradation of catechol produced from both phenol and salicylate. In the wild strain, a certain portion of catechol (maybe most of the portion produced from salicylate) seemed to be converted into 2-HMS and the following intermediates on the *meta*-cleavage pathway. Thus, phenol degradation by the GEM might be faster than that of the wild strain.

These inferences concerning catabolic gene expression in both strains were confirmed by the assay of catechol 1,2-oxygenase (C120) which is the key enzyme of the *ortho*-cleavage pathway and C230 which, on the other hand, is that of a *meta*-cleavage pathway. The assay was performed by using the cell-free crude extract prepared from the culture broth which was sampled periodically during the degradation tests, and the results are shown in Figure 5.6.

Interestingly, in the simultaneous degradation of phenol and salicylate (in phenol-salicylate medium), the GEM grew faster than the wild strain, the opposite outcome of the degradation test using only salicylate-containing medium. Moreover, complete removal of either substrate (phenol or salicylate) by the GEM was completed sooner than by the wild strain.

5.2.4 Enhanced Activity

The results of degradation tests in phenol media, salicylate media, and phenol-salicylate media by the GEM, *P. putida* PpG1064 (pHF400), and the wild strain, *P. putida* PpG1064 (NAH), are compared in Table 5.4. The results shown in Table 5.4 are expressed by the specific growth rate (μ), and substrate removal rates (K) defined as follows.

$$\mu = (dX/dt)/X \tag{5.1}$$

$$K = (-dS/dt)/S \tag{5.2}$$

where X is the cell concentration which was measured as OD_{600}, S is the concentration of phenol or salicylate, and t is time. K is equal to the first-order constant for substrate removal. The μ and K values in the table were

TABLE 5.4. Specific growth rates and substrate degradation rates in P. putida *PpG1064 (pHF400) and* P. putida *PpG1064 (NAH).*

		Medium/Substrate			
				Phenol-Salicylate	
Strain	Rate[a]	Salicylate	Phenol	Salicylate	Phenol
PpG1064 (NAH)	μ	0.173	0.163	0.172	
	δ	0.254	0.562	0.190	0.455
PpG1064 (pHF400)	μ	0.166	0.167	0.198	
	δ	0.210	0.574	0.437	0.614

[a]μ: specific growth rate (1/hr), the slope of the growth curve in semi-logarithmic plot.
δ: substrate degradation rate (1/hr), the slope of the degradation curve in semi-logarithmic plot.

TABLE 5.5. C120 and C230 activity in P. putida *PpG1064 (pHF400) and* P. putida *PpG1064 (NAH).*

		Medium		
Strain	Enzyme	Salicylate	Phenol	Phenol-Salicylate
PpG1064 (NAH)	C120	778	155	383
	C230	12	2,060	324
PpG1064 (pHF400)	C120	1,280	1,420	1,310
	C230	–	–	–

determined as the mean values observed in the logarithmic growth phase, and so they may be maximum values for each test.

In addition, the maximum activities of C120 and C230 in the tests are summarized in Table 5.5.

As shown in Table 5.4, the rate of salicylate degradation of the GEM was slightly lower than that of the wild strain, although the phenol degradation rate of the GEM was equal to or a little higher than that of the wild strain. Therefore, it is apparent that the GEM, *P. putida* PpG1064 (pHF400), is not inferior to the wild strain, *P. putida* PpG1064 (NAH), in its ability to degrade phenol and salicylate in single substrate media. Therefore, the GEM may be fully capable of degrading salicylate, even though it has to use the *ortho*-cleavage pathway of the host.

Furthermore, the GEM showed apparently higher specific growth and substrate removal rates than the wild strain in the simultaneous degradation of phenol and salicylate. Particularly, the salicylate degradation rate of the GEM was 2.3 times higher than that of the wild strain. Thus, the degrading activities for phenol and salicylate were enhanced by genetic manipulation. This enhancement of degrading activities may depend on the differences in the organization and transcriptional control of corresponding catabolic genes between the GEM and the wild strain. The genetic differences are as follows:

- In the GEM, the *nahG* gene carried on the recombinant plasmid pHF400 exists at a high copy number (approximately 10–20 copies per cell), while only one or a few copies of the *nahG* gene per cell are contained in the wild strain, because NAH is a naturally occurring low copy number plasmid.
- As the *nahG* gene is transcribed constitutively from the *Km* promotor on pHF400, salicylate hydroxylase is produced constitutively in the GEM. On the other hand, the expression of *nahG* is positively regulated, with salicylate as an inducer in the wild strain.

- The catabolic pathway for catechol in the GEM is the *ortho*-cleavage pathway only, whereas the wild strain possesses both the *ortho*- and *meta*-cleavage pathways. The wild strain alternatively uses *ortho*-cleavage for phenol degradation and *meta*-cleavage for salicylate degradation in the single substrate media, as shown in Table 5.5. (The C12O and C23O activities were expressed alternatively.) However, whether the *ortho*-cleavage pathway or *meta*-cleavage pathway is mainly used for degrading phenol or salicylate could not be ascertained when both substrates were degraded simultaneously (in phenol-salicylate medium).

Among these three, the latter is considered the most significant factor affecting growth and substrate degradation properties, judging from the experimental data. When phenol and salicylate were degraded simultaneously, the sum of C12O and C23O activities in the wild strain, *P. putida* PpG1064 (NAH), was approximately half the C12O activity expressed in the GEM, *P. putida* PpG1064 (pHF400), suggesting faster degradation of catechol in the GEM than in the wild strain. The faster degradation of catechol in the GEM may also be associated with superior growth, or phenol and salicylate degradation. The C12O activity of the GEM in phenol-salicylate medium was as high as that in phenol or salicylate medium. In the wild strain, however, the sum of C12O and C23O activities in phenol-salicylate medium was lower than the C12O activity in phenol medium or the C23O activity in salicylate medium. This suggests a disadvantage to the existence of both *ortho*- and *meta*-cleavage pathways in simultaneous degradation.

5.2.5 Conclusions and Considerations

In this study, *P. putida* PpG1064, a plasmid-free derivative strain of the wild strain, *P. putida* PpG1064 (NAH), was used as the host of the recombinant plasmid pHF400 to create the GEM capable of degrading salicylate. We have found a number of merits to this approach.

First, the catechol catabolic pathway coded on the chromosome of the host could be complemented by the plasmid-coded *nahG* gene to establish a complete catabolic pathway for salicylate. This indicates an ability to extend the catabolic range of a bacterial host (one which can degrade an aromatic compound) by simple genetic manipulation. Since bacteria capable of degrading aromatic compounds generally have a catechol pathway, the addition of only one or a few gene(s) coding for the enzyme(s) responsible for the transformation of the target aromatics into catechol can lead to complete degradation through the catechol pathway possessed by the host.

Second, the plasmid burden may be reduced by replacing the naturally occurring catabolic plasmid with the recombinant plasmid to eliminate the unnecessary part for salicylate degradation. For example, it was considered that the plasmid burden by the NAH plasmid might be greater than that of the recombinant plasmid pHF400, since the former (83 kb) is much greater than the latter (15 kb) in size. However, this effect was not seen clearly in this study, although the growth and phenol degradation rates in phenol medium were slightly higher in the GEM than in the wild strain. This might depend on the difference in the copy number between both plasmids.

The GEM may show higher degradation activities than the wild strain when degrading multiple substrates simultaneously, as was observed in this study. Further, in the simultaneous degradation of various aromatic compounds, catechol degradation by the *ortho*-cleavage pathway solely seems to be preferable to that by both the *ortho*- and *meta*-cleavage pathways. This advantage of the GEM is very applicable to wastewater treatment.

5.3 AMPLIFICATION OF CATABOLIC PATHWAY

5.3.1 Concept

The ultimate purpose of the removal of xenobiotic compounds in the wastewater treatment is complete degradation. In a sense, *complete degradation* means the transformation of target compounds into TCA cycle intermediates such as citrate, succinate, acetyl-CoA, etc., which can be utilized by various common bacteria. In general, xenobiotic or recalcitrant compounds are completely metabolized through a series of biotransformation steps, and many catabolic enzymes (and corresponding genes) participate in complete degradation. For example, the phenol degradation pathway found in *P. putida* BH is considered to be composed of the following six enzymatic steps (see Section 4.3; Figure 4.10).

1st : phenol to catechol (catalyzed by phenol hydroxylase) (5.3.a)

2nd : catechol to 2HMS (catalyzed by C23O) (5.3.b)

3rd : 2HMS to 2-oxopent-4-enoate (catalyzed by 2HMS hydrolase [2HMSH]) (5.3.c)

4th : 2-oxopent-4-enoate to 4-hydroxy-2-oxovalerate (catalyzed by 2-oxopent-4-enoate hydratase [OEH]) (5.3.d)

5th : 4-hydroxy-2-oxovalerate to pyruvate and acetaldehyde (catalyzed by 4-hydroxy-2-oxovalerate aldolase [HOA]) (5.3.e)

6th : acetaldehyde to acetyl-CoA
(catalyzed by acetaldehyde dehydrogenase [ADA]) (5.3.f)

The third step (5.3.c) and the following three steps are used alternatively to transform 2HMS to 2-oxopent-4-eonate. When the following pathway is selected, the phenol degradation pathway in *P. putida* BH can be divided into eight enzymatic reactions.

3–1st : 2HMS to 4-oxalocrotonate, enol form
(catalyzed by 2HMS dehydrogenase [2HMSD]) (5.3.g)

3–2nd : 4-oxalocrotonate, enol form to keto form
(catalyzed by 4-oxalocrotonate isomerase [4OI]) (5.3.h)

3–3rd : 4-oxalocrotonate, keto form to 2-oxopent-4-enoate
(catalyzed by 4-oxalocrotonate decarboxylase [4OD]) (5.3.i)

The rate of complete degradation of xenobiotic compounds (in which the pathway is composed of multiple reactions) is governed by the rate-determining steps on the pathway. Therefore, such reaction(s) should be determined, and the expression of corresponding catabolic enzyme(s) amplified, in order to enhance the degrading activity of the existing microorganisms.

As for the phenol degradation, of course, the first step (5.3.a), which is the transformation of phenol into catechol, should be the most probable rate-determining step of the metabolic pathway. However, since accumulation of a considerable amount of catechol in the culture medium during phenol decomposition was reported in several studies which dealt with the microbial degradation of phenol (Evans, 1974; Itoh and Fujiwara, 1979; Nei et al., 1974), it has been suggested that the reaction catalyzed by catechol oxygenase (the step catalyzed by either C12O or C23O—the reaction [5.3.b]—when phenol is degraded by the *meta*-cleavage) may be another rate-determining step in addition to the first step. *P. putida* BH also occasionally accumulates only a small amount of catechol during phenol degradation. The occurrence of this phenomenon is considered to depend on the culture conditions. Thus, the conversion of phenol to catechol, and that of catechol to 2HMS, are the most likely rate-determining steps governing phenol degradation.

In this section, on the assumption that the conversion steps (5.3.a) and/or (5.3.b) can be the rate-determining steps on the phenol catabolic pathway of *P. putida* BH in a certain condition, endeavors were made to enhance the phenol degradation activity by amplifying the corresponding genes, *pheA* and *pheB*, which code for phenol hydroxylase and C23O, respectively, by simple genetic manipulation, and their effects were evaluated.

5.3.2 Amplification of Phenol Hydroxylase Gene

First, the *pheA* gene cloned from the chromosome of *P. putida* BH was reintroduced into the parental strain *P. putida* BH as the recombinant plasmid pS4-92 for amplifying a putative rate-determining step (5.3.a) in the phenol degradative pathway.

As described in Section 4.3, *Sal*I-digested total genomic DNA of *P. putida* BH was cloned into the cosmid vector pVK100 to produce a gene library in *E. coli* HB101, and seven clones positive for phenol catabolic genes, at least including the *pheA* gene, were identified among a total of approximately 2,300 clones. The hybrid cosmid present in one such positive clone, designated pS4-92, was used in this study. It contained approximately a 21.9-kb DNA fragment from *P. putida* BH, and, when transferred into *P. putida* KT2440, which cannot grow on phenol but can grow on catechol, conferred the ability to grow on phenol. Therefore, it was confirmed that pS4-92 carried the *pheA* gene. However, since both *E. coli* HB101 (pS4-92) and *P. putida* KT2440 (pS4-92) lacked detectable C230 and other *meta*-cleavage enzyme activities, the pS4-92 seemed not to contain only one gene coding for a phenol catabolic enzyme, phenol hydroxylase. On the other hand, there was some evidence for the existence of the gene encoding the positive regulator (*pheR*) of the phenol catabolic pathway (*pheA*) of *P. putida* BH on pS4-92.

The hybrid cosmid, pS4-92, was transferred from *E. coli* HB101 (pS4-92) into the parental strain, *P. putida* BH, by triparental conjugation with a helper strain, *E. coli* HB101 (pRK2013). One of the transconjugants, designated *P. putida* BH (pS4-92), was selected for evaluating the effects of this gene amplification on phenol degradation. The GEM, *P. putida* BH (pS4-92), possessed a few or several copies of the *pheA* and *pheR* genes on pS4-92 and the chromosome, and only one copy of each gene for the *meta*-cleavage pathway of catechol, while the parental strain possessed only one copy of all genes on its chromosome.

Phenol removal tests of the GEM, *P. putida* BH (pS4-92), and the parental strain, *P. putida* BH, were carried out by using phenol medium (minimal salts medium containing phenol as a sole carbon and energy source at the concentration of 400 mg/L) for comparison of their phenol degrading properties. Both strains were precultured in LB broth supplemented with phenol at 400 mg/L for inducing the phenol hydroxylase production (expression of the *pheA* gene coded on both the chromosome and the recombinant plasmid), and collected by centrifugation. Washed cell pellets (with cold 50-mM phosphate buffer) were suspended in 100 mL of phenol medium at a definite cell density (OD_{600} = 1.0, equal to approximately 350 mg-dry weight/L), and the cell suspension contained in a 300-mL Erlenmeyer flask

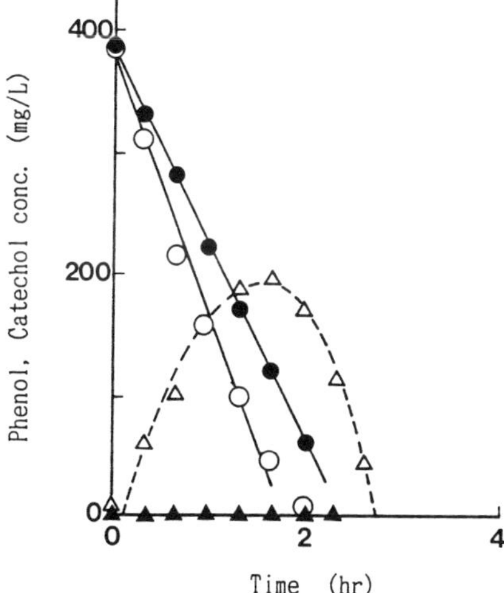

FIGURE 5.7. Phenol degradation by *P. putida* BH and *P. putida* BH (pS4-92). ●, phenol removal and ▲, catechol accumulation by BH; ○, phenol removal and △, catechol accumulation by BH (pS4-92).

was shaken on a rotary shaker (160 rpm) at 30°C. Phenol and a key intermediate, catechol, in the medium were monitored periodically by HPLC assay.

The results are shown in Figure 5.7. As shown in the figure, the GEM, *P. putida* BH (pS4-92), removed phenol in the medium faster than the parent strain, *P. putida* BH. From the data obtained here, a specific phenol removal rate (K) defined as follows, was calculated for both the GEM and the wild strain.

$$K = (-dPhe/dt)/X \tag{5.4}$$

where Phe is the phenol concentration, X is the cell concentration, and t is time. As the cell concentration (X) was maintained at a constant level in this experiment (the cell density was so high that cell growth during the experimental period was negligible) and the slope which appeared in Figure 5.7 expresses ($-dPhe/dt$), the specific phenol removal rate (K) could be determined as the slope divided by the initial cell concentration. The K values calculated in this manner were 0.682 and 0.465 g of phenol removed/g-cell/hour on the GEM and the parental strain respectively. The K value of the GEM was about 1.5 times higher than that of the parental strain. Therefore, it appeared that phenol removal activity was enhanced by amplifying the *pheA* gene with the hybrid cosmid, pS4-92.

However, although the GEM transformed phenol into catechol faster than the parent strain, it accumulated a considerable amount of catechol in the medium, while no detectable amount of catechol was accumulated by the parental strain. For example, when approximately 230 mg/L of phenol was removed by the GEM (at 1 hour after the phenol removal test started), catechol was accumulated at 160 mg/L in the medium, indicating that only 93 mg/L of phenol was degraded into 2HMS or other downstream intermediates such as 2-oxopent-4-enoate, 4-hydroxy-2-oxovalerate, pyruvate, acetaldehyde, etc., and, therefore, that at least 137 mg/L of phenol converted into catechol was not further degraded at that time. On the other hand, about 130 mg/L of phenol was transformed by the parental strain into the intermediates degraded further than 2HMS at that time. Such accumulation of the intermediate by the GEM suggests that the complete degradation of phenol could not be enhanced by manipulation of the *PheA* gene, although the conversion rate of phenol to catechol increased.

Moreover, in this experiment, even when both phenol and catechol were removed by the parental strain—suggesting complete degradation—catechol was not degraded and remained at approximately 50 mg/L. This indicates that the catechol degradation activity of the host, *P. putida* BH, was suppressed by amplifying the expression of the *pheA* gene with the recombinant plasmid, pS4-92. This suppression may be due to the enhanced expression of the *pheA* gene causing a rapid decrease of phenol in the medium, which was the inducer of the catechol-degrading enzymes such as C230, 2HMSH, OEH, etc., which were coded only on the chromosome of the host.

From these results of the phenol removal tests, it is thought that genetic manipulation on a rate-determining step in the catabolic pathway of xenobiotic compounds may generate other rate-determining steps, "secondary rate-determining steps," and that it may not always lead to the enhancement of complete degradation.

5.3.3 Amplification of C230 Gene

The second putative rate-determining step on the phenol catabolic pathway of *P. putida* BH is the conversion of catechol to 2HMS (5.3.b) catalyzed by C230 which is specified by the *pheB* gene. Moreover, it was shown that this reaction actually became a considerable rate-determining step in the GEM, *P. putida* BH (pS4-92), which possessed the genetically modified phenol catabolic pathway on which the reaction catalyzed by phenol hydroxylase (5.3.a) was amplified. In this section, another GEM, of which the phenol catabolic pathway was amplified at the step catalyzed by C230 (5.3.b), was constructed by reintroducing the recombinant plasmid, pBH500, carrying the cloned *pheB* gene into the parental strain, *P. putida* BH.

The recombinant plasmid pBH500 was constructed by inserting the 5.65-kb DNA fragment containing C230 encoding *pheB* gene cloned from the chromosome of *P. putida* BH into a broad host range vector plasmid, pKT230. The details of the construction of this recombinant plasmid were described in Section 5.1. The pBH500 did not have any other phenol catabolic genes but the *pheB* gene. The *pheB* gene coded on pBH500 is constitutively transcribed from the *Km* promotor of the vector, as the cloned DNA fragment did not contain any *P. putida* BH-derived promotor region.

The pBH500 was transferred into *P. putida* BH by triparental conjugation from donor *E. coli* C600 (pBH500) by the mediation of a mobilizer strain of *E. coli* C600 (RP4), and consequently the GEM *P. putida* BH (pBH500) was constructed. The GEM *P. putida* BH (pBH500) seemed to possess approximately 15–20 copies of the *pheB* gene on the recombinant plasmid pBH500, and only one copy on the chromosome, while the parental strain possessed only one copy on its chromosome. Each strain contained one copy of another phenol catabolic genes on the chromosome. This GEM breeding technique for amplifying the second step reaction (5.3.b) of the phenol degradation pathway in *P. putida* BH is diagramed in Figure 5.8.

To evaluate the degree of the amplification of the gene expression of *pheB*, the C230 activities of *P. putida* BH (pBH500) grown in LB broth containing phenol at different concentrations (0, 100, 200, 300, and 400 mg/L) were assayed and compared with those of the parental strain, *P. putida* BH. Results are presented in Table 5.6.

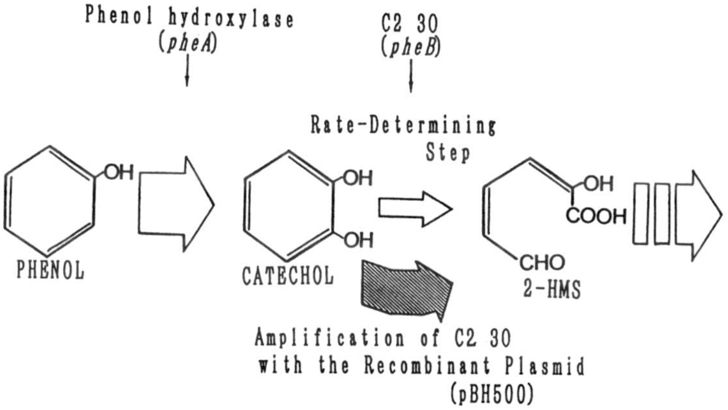

FIGURE 5.8. Concept of amplification of C230-catalyzing step in *P. putida* BH.

TABLE 5.6. C230 activity of P. putida *BH and* P. putida *BH (pBH500).*

	C230 activity (Units/mg-protein)			
Phenol conc.	BH	BH (pBH500)	difference[a]	ratio[a]
None	<1	23.9	23.9	–
100 mg/L	29.2	49.0	19.8	1.68
200 mg/L	54.4	75.8	21.4	1.39
300 mg/L	62.8	80.6	17.8	1.28
400 mg/L	100.4	121.2	20.8	1.21

[a]C230 activity of BH (pBH500) – C230 activity of BH.
[b]C230 activity of BH (pBH500)/C230 activity of BH.
Reprinted from, *Water Research, Vol. 27,* Fujita, Ike and Kamiya, "Accelerated Phenol Removal by Amplifying the Gene Expression with a Recombinant Plasmid Encoding Catechol-2,3-Oxygenase," p. 11, 1993, with kind permission from Pergamon Press, Ltd., Headington Hill Hall, Oxford OX3 0BW, UK.

As shown in the table, the parental (wild) strain expressed the *pheB* gene, which was coded on the chromosome, and showed C230 activity only when induced with phenol, its activity tended to be proportional to phenol concentration. On the other hand, genetically engineered *P. putida* BH (pBH500) consistently expressed the *pheB* gene coded on the recombinant plasmid, and exhibited C230 activity constitutively regardless of whether or not phenol was added to the culture medium. However, when induced with phenol, the C230 activity of the GEM was enhanced, perhaps due to the "plus effect" by the induced expression of the *pheB* gene on the chromosome; the expression of the *pheB* gene coded on the chromosome might be added to that of the recombinant plasmid. The GEM showed higher C230 activity than did *P. putida* BH regardless of the phenol concentration used for the cultivation of the strains. In any induction level of phenol, the difference in the C230 activity between the GEM and the parental strain was approximately 20 units/mg-protein, regardless of phenol concentration. Therefore, as the phenol concentration of the growth medium increased, the effect of the gene dosage for C230 production decreased (see the "ratio" column in Table 5.6).

Phenol removal tests of the GEM, *P. putida* BH (pBH500), and the wild strain, *P. putida* BH, were performed by batch culture as follows. One milliliter of overnight culture of each strain in LB broth was inoculated into 100 mL of new LB broth in 300-mL Erlenmeyer flasks and cultivated on a rotary shaker (160 rpm) at 30°C for about 18 hours. Then, 1 mL of this culture was inoculated into 200 mL of phenol medium in a 500-mL Erlenmeyer flask and the batch cultivation was carried out in the same manner. During cell growth, (OD_{600} cell density), phenol, catechol, 2HMS, and TOC concentrations were measured periodically.

Phenol concentrations contained in the test medium were set at four different levels (100, 200, 300, 400 mg/L). Typical time courses of cell growth and substrate degradation (phenol, catechol, 2HMS, and TOC concentrations) of the GEM, *P. putida* BH (pBH500), and the parental strain, *P. putida* BH, are shown in Figure 5.9. Figure 5.9(a) and (b) show the results of the tests using phenol medium containing 100 and 300 mg/L phenol, respectively. Although both *P. putida* BH (pBH500) and *P. putida* BH degraded phenol completely and efficiently in the batch cultures, their growth and substrate removal profiles were different, and the differences depended upon the initial phenol concentration of the medium.

At the initial phenol concentration of 100 mg/L, the GEM grew and removed phenol and TOC much faster than the parental strain [Figure 5.9(a)]. On the other hand, when phenol was contained in the test medium at initial concentrations of 200 and 300 mg/L, both strains grew and removed phenol at approximately the same rates. Although the differences in cell growth and phenol removal between the GEM and the parental strain were not observed clearly in these conditions, the GEM showed a tendency to grow a little faster than the parental strain in the early cultivation period [Figure 5.9(b)]. However, at an initial phenol concentration of 400 mg/L, the parental strain grew and removed phenol slightly faster than the GEM. As for the intermediates, catechol concentration was kept under the detection limit of HPLC assay (less than 1–3 mg/L) in all tests for both strains, suggesting no accumulation of catechol, while 2-HMS was produced and further degraded at slightly higher rates by the GEM than by the parental strain, although accumulated concentrations were very low (below 1 mg/L).

It was generally demonstrated that growth and substrate degradation in the batch culture were divided into two nondistinct logarithmic phases—one phase may be the induction phase during which the corresponding degradative enzymes or activities are not induced fully, and the other phase may be the postinduction phase. Since this observation fit the results obtained in these batch studies of *P. putida* BH and *P. putida* BH (pBH500), the growth and phenol removal properties of both strains were evaluated for both phases.

The specific growth rates (μ), defined in (2.6), for both phases were calculated from the slopes of the semi-logarithmic plot of the growth curves, and these values were used for the evaluation of the growth characteristics of both the GEM and the parental strain. The specific growth rate for the induction phase, and that for the postinduction phase, are expressed as μ_1 and μ_2.

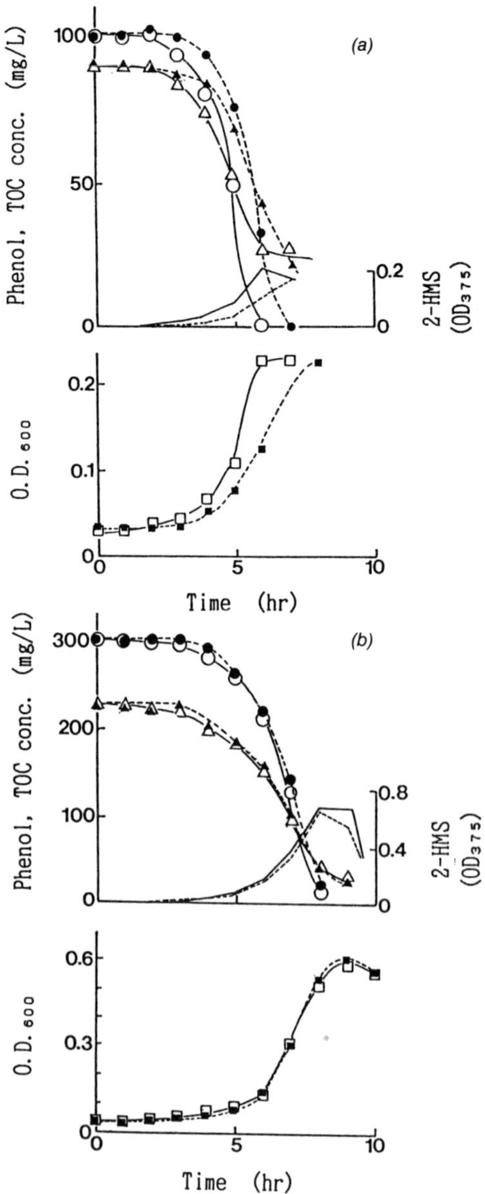

FIGURE 5.9. Phenol degradation by *P. putida* BH and *P. putida* BH (pBH500). (a) Initial phenol concentration = 100 mg/L. (b) Initial phenol concentration = 300 mg/L. ●, phenol; ▲, TOC removal; - - -, 2-HMS accumulation by and ■, cell growth of BH; ○, phenol; △, TOC removal; —, 2-HMS accumulation by and □, cell growth of BH (pBH500). Reprinted from *Water Research, Vol. 27,* Fujita, Ike and Kamiya, "Accelerated Phenol Removal by Amplifying the Gene Expression with a Recombinant Plasmid Encoding Catechol-2,3-Oxygenase," p. 11, 1993, with kind permission from Pergamon Press Ltd., Headington Hill Hall, Oxford OX3 0BW, UK.

$$\mu_1, \mu_2 = (dX/dt)/X \; \mu_1, \text{ induction phase; } \mu_2, \text{ post-induction phase} \tag{5.5}$$

where X is the cell concentration and t is time.

On the other hand, the phenol and TOC removal rates were evaluated by the Kp and Kt values defined as follows, and they were also calculated from the slopes of the semilogarithmic plots of phenol and TOC degradation curves in the same manner for $\mu - Kp_1$ and Kt_1 for the induction phase, and Kp_2 and Kt_2 for the postinduction phase.

$$Kp_1, Kp_2 = (-dPhe/dt)/Phe$$
$$KP_1, \text{ induction phase; } Kp_2 \text{ postinduction phase} \tag{5.6}$$

$$Kt_1, Kt_2 = (-dTOC/dt)/TOC$$
$$Kt_1, \text{ induction phase; } Kt_2, \text{ postinduction phase} \tag{5.7}$$

where *Phe* is the phenol concentration and *TOC* is the TOC concentration. The μ, Kp, and Kt values calculated from the data were plotted against the initial phenol concentration of the medium and shown in Figure 5.10(a), (b), and (c).

As shown in the figures, μ_1, Kp_1, and Kt_1 of the GEM, *P. putida* BH (pBH500), were higher than those of the parental strain, *P. putida* BH, when grown at lower phenol concentrations. Other parameters of the GEM were, however, nearly equal to or slightly lower than those of the parental strain as a whole. From these results, it is apparent that accelerated phenol removal by the amplification of the expression of the *pheB* gene was especially demonstrated in the induction phase of the batch cultivation at relatively low phenol concentration.

Therefore, it may be concluded that the C23O catalyzing reaction (5.3.b) was the rate-determining step of phenol degradation in *P. putida* BH in the induction phase and/or at relatively low phenol concentration. In the experiments here, catechol was not detected by the HPLC assay, but it is considered that a very small amount of catechol was actually accumulated, and that, therefore, the conversion of catechol into 2HMS (5.3.b) was accelerated slightly by the amplification of the *pheB* gene, further enhancing phenol degradation. That the TOC removal rate was higher in the GEM than in the parental strain under such conditions can be understood easily by such inferences. For example, whereas the phenol removal rate—the conversion rate of phenol into catechol (5.3.a)—was enhanced by the genetic manipulation of *pheB*, this may be difficult to elucidate because in theory this conversion step (5.3.a) seems not to be affected by amplifying the *pheB* gene.

Even at relatively higher phenol concentrations, the effects of gene

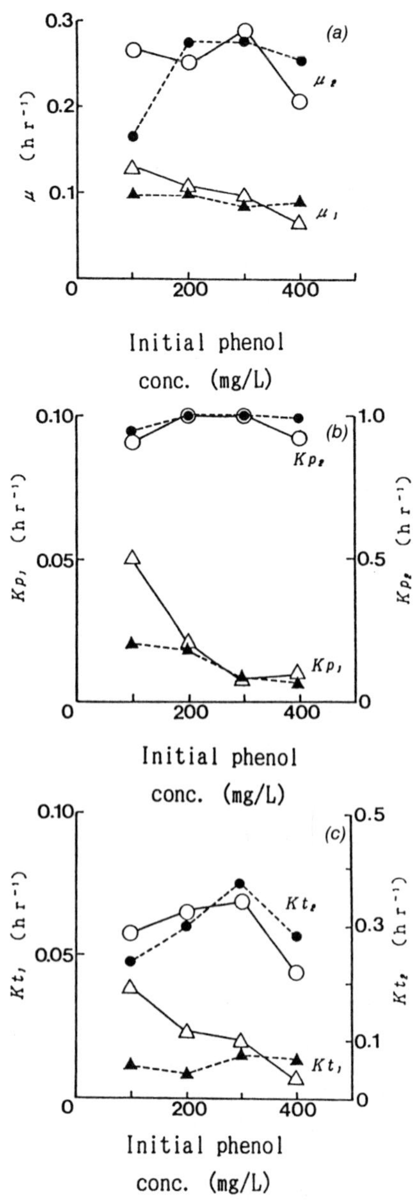

FIGURE 5.10. Effect of initial phenol concentration on the growth, phenol removal, and TOC removal rates of *P. putida* BH and *P. putida* BH (pBH500). (a) Effect on specific growth rate (μ). (b) Effect on phenol removal rate (*Kp*). (c) Effect on TOC removal rate (*Kt*). ▲, μ_1, Kp_1; Kt_1 and ●, μ_2, Kp_2, Kt_2 of BH; △, μ_1, Kp_1, Kt_1 and ○, μ_2, Kp_2, Kt_2 of BH (pBH500). (Reprinted from *Water Research, Vol. 2,* Fujita, Ike and Kamiya, "Accelerated Phenol Removal by Amplifying the Gene Expression with a Recombinant Plasmid Encoding Catechol-2,3-Oxygenase," p. 12, 1993, with kind permission from Pergamon Press, Ltd., Headington Hill Hall, Oxford OX3 0BW, UK.

amplification were not observed clearly. Possible reasons for little or no enhancement of phenol degradation activity by the gene amplification of *pheB* under such condition may be:

- The conversion reaction (5.3.b) catalyzed by C230 was really not the rate-determining step on the phenol catabolic pathway of *P. pituda* BH at higher phenol concentrations.
- The effect of the amplification of gene expression on the enhancement of the C230 activity was reduced at higher phenol concentrations, as shown in Table 5.6.
- The plasmid burden of the pBH500 affected the growth of the GEM considerably more under such conditions.

In the actual wastewater treatment process, the concentration of xenobiotic compounds contained in the influent is relatively low and fluctuates considerably. Therefore, the most desirable characteristics of degrading microorganisms for wastewater treatment are considered to be high growth rate and degradation activity at low substrate concentration, and rapid induction of degradative enzymes. From this viewpoint, *P. putida* BH (pBH500) is anticipated to have a wide scope in effective phenol removal in wastewater treatment.

5.4 CONCLUSIONS AND CONSIDERATIONS

For improving phenol degradation activity of the wild bacterial strain, *P. putida* BH, the enzymatic reactions (5.3.a) catalyzed by phenolhydroxylase (the transformation of phenol into catechol) and (5.3.b) catalyzed by C230 (the transformation of catechol into 2HMS)—each of which was presumed to be a rate-determining step on the phenol catabolic pathway—were amplified by introducing the recombinant plasmids carrying the phenol hydroxylase-encoding *pheA* and the C230-encoding *pheB* gene, and the resultant GEMs, *P. putida* BH (pS4-92) and *P. putida* BH (pBH500), respectively, were constructed.

Although *P. putida* BH (pS4-92) could transform phenol into catechol much faster than the parental strain, the rate of complete degradation (further degradation of catechol) was not improved by gene amplification of *pheA*. During phenol degradation, *P. putida* BH (pS4-92) accumulated a considerable amount of catechol in the medium, suggesting that the conversion of the catechol into 2HMS became a rate-determining step on the phenol catabolic pathway of the GEM as the result of genetic manipulation.

On the other hand, *P. putida* BH (pBH500) grew and removed both phenol and TOC much faster than the parental strain in the minimal salts medium containing phenol at relatively low concentration. However, at higher phenol concentration, differences in cell growth and phenol removal between the GEM and the parental strain were not observed clearly, while the GEM showed a tendency to grow a little faster than the parental strain in the early cultivation period. Thus, accelerated phenol removal by the amplification of the *pheB* expression was especially demonstrated in the induction phase of batch cultivation at relatively low phenol concentration.

From these results, it is concluded that the genetic amplification of a rate-determining step on the catabolic pathway which consists of multiple enzymatic reactions, may result in either success or failure in the improvement of the degradation rate of target compounds. Possible reasons for failure may be:

- The amplified reaction is not always the rate-determining step on the pathway; it seems to be determined by environmental conditions such as substrate concentration, temperature, etc.
- The amplification of a rate-determining step may generate another or secondary rate-determining step, resulting in the accumulation of intermediates on the pathway. Thus, the rate of complete degradation cannot be improved.
- Genetic manipulation may have adverse effects on the basic metabolism/anabolism activities of the host, such as the plasmid burden, for example. It may counterbalance the enhanced degradation activity.

The first problem suggests that each reaction on the catabolic pathway can be a rate-determining step under a certain environmental condition, and may be solved by fully investigating the conversion rate of each reaction under various conditions. Such investigation should help in the rational design of gene amplification and GEMs to enhance, case by case, the degradation rate of xenobiotic compounds in wastewater treatment.

The second problem may be related to the degree of amplification of the rate-determining step. A simple countermeasure for dissolving the secondary rate-determining step is to further amplify this reaction; however this may result in the occurrence of a third rate-determining step. Although another presumable countermeasure is to amplify not just one or a few reactions but the whole pathway, such genetic manipulation may cause adverse effects (the third problem) because of too high a dose of introduced genes on the host.

Therefore, various types of the GEMs as described in this section should be constructed, and their growth and substrate removal properties evaluated, to obtain suitable GEMs applicable to wastewater treatment.

5.5 REFERENCES

Bagdasarian, M., R. Luze, B. Rückert, F. C. H. Franklin, M. M. Bagdasarian, J. Frey and K. N. Timmis. 1981. "Specific-Purpose Plasmid Cloning Vectors, II. Broad Host Range, High Copy Number, RSF1010-Derived Vectors and Host-Vector System for Gene Cloning in *Pseudomonas*," *Gene*, 16:237–247.

Boyer, H. W. and Roulland-Dussoix, D. 1969. "A Complementation Analysis of the Restriction and Modification of DNA in *Escherichia coli*," *J. Mol. Biol.*, 41:459–472.

Dunn, N. W. and I. C. Gunsalus. 1973. "Transmissable Plasmid Coding Early Enzymes of Naphthalene Oxidation in *Pseudomonas putida*," *J. Bacteriol.*, 114:974–979.

Evans, W. C. 1974. "Oxidation of Phenol and Benzoic Acid by Some Soil Bacteria," *Biochem. J.*, 41:373–382.

Fewson, C. A. 1981. "Biodegradation of Aromatics with Industrial Relevance," in *Microbial Degradation of Xenobiotics and Recalcitrant Compounds*, T. Leisinger, R. Hütter, A. M. Cook and J. Nüesch, eds., London: Academic Press Inc., pp. 141–179.

Fujita, M., T. Kamiya, M. Ike., Y. Kawagoshi and N. Shinohara. 1991. "Catechol 2,3-Oxygenese Production by Genetically Engineered *Escherichia coli* and its Application to Catechol Determination," *Wld. J. Microbiol. Biotechnol.*, 7:407–414.

Harayama, S., M. Rekik and K. N. Timmis. 1986. "Genetic Analysis of a Relaxed Substrate Specificity Aromatic Ring Dioxygenase, Toulene 1,2-Dioxygenase, Encoded by TOL Plasmid pWWO of *Pseudomonas putida*," *Mol. Gen. Genet.*, 202:226–234.

Itoh, M. and N. Fujiwara. 1979. "Degradation of Phenol by Yeasts," *J. Ferment. Technol.*, 57:421–428 (In Japanese).

Lehrbach, P. R., J. Zeyer, W. Reineke, H.-J. Knackmuss and K. N. Timmis. 1984. "Enzyme Recruitment *in Vitro*: Use of Cloned Gene to Extend the Range of Haloaromatics Degraded by *Pseudomonas* sp. Strain B13," *J. Bacteriol.*, 58:1025–1032.

McClure, N. C., J. C. Fry and A. J. Weightman. 1991. "Genetic Engineering for Wastewater Treatment," *J. IWEM.*, 5:608–616.

Nakazawa, T. and T. Yokota. 1973. "Benzoate Metabolism in *Pseudomonas putida (arvilla)* mt-2, Demonstration of Two Benzoate Pathway," *J. Bacteriol.*, 115:262–267.

Nei, N., Y. Tanaka and N. Takata. 1974. "Production of Catechol from Phenol by a Yeast," *J. Ferment. Technol.*, 52:28–34 (In Japanese).

Nozaki, M. 1970. "Metapyrocatechase (*Pseudomonas*)," *Method in Enzymol.*, 17A:522–525.

Rojo, F., D. H. Pieper, K.-H. Engesser, H.-J. Knackmuss and K. N. Timmis. 1987. "Assemblage of Ortho Cleavage Route for Simultaneous Degradation of Chloro- and Methylaromatics," *Science*, 238:1395–1398.

CHAPTER 6

Genetic Stability of Genetically Engineered Microorganisms and Wastewater Treatment

The purpose of wastewater treatment is to constantly maintain pollutants contained in effluent at lower than a defined water quality standard; the stability of the treatment is very significant. In the application of GEMs to wastewater treatment, the gene expression or activity of the introduced GEMs should be maintained at a sufficiently high level of stability so as to function fully in actual wastewater treatment. The most basic problem with GEMs is their genetic stability—the stability of the recombinant gene itself and its expression mechanisms in the host.

This chapter will deal with the genetic stability problems of GEMs from the viewpoint of their application in the wastewater treatment processes. Usually, genetic stability problems first become evident in the GEM breeding step. For example, the introduced foreign DNA fragments may be degraded or digested completely or partly by the endonucleases produced in the host strain, and therefore the desired GEM cannot be obtained. These problems are not covered in this chapter, which deals only with the stability of completely constructed GEMs with very stable genotypes and phenotypes, such as those described in Chapter 5. Special emphasis is given to the stability of recombinant plasmids, which are the most common GEM breeding materials. Properties of GEMs suitable for wastewater treatment are discussed vis a vis results of stability tests of recombinant plasmids in a variety of host strains.

6.1 STABILITY OF RECOMBINANT PLASMIDS

6.1.1 Instability Phenomena

Since the construction of GEMs is usually accomplished by introducing recombinant plasmids containing cloned genes into host bacterial cells, recombinant plasmids are involved in genetic stability problems. In general, *plasmid stability* means the ability of GEMs to maintain their recombinant plasmids unchanged during growth, i.e., stable expression of their phenotypic characteristics.

Several experimental studies have described GEMs harboring unstable or relatively stable plasmids in continuous cultures simulating the microbial production of valuable materials by the GEMs (Jones et al., 1980; Mizutani et al., 1985; Noack et al., 1981; Weber and San, 1990; Koizumi et al., 1985). The maintenance and stability of the catabolic plasmids were also studied (Duetz and van Andel, 1991; Williams et al., 1988). Further, the stability of plasmid expression and maintenance during long-term starvation-survival of various bacterial strains in well water, groundwater, etc. were investigated to assess possible biohazards caused by GEMs (Caldwell et al., 1989; Jain et al., 1987).

In such research, it has been generally observed that plasmid-free cells (segregants) appear after a certain lag phase in the continuous culture of plasmid-harboring strains with nonselective media. Subsequently, the plasmid-free cells usually increased as the cultivation proceeded.

For example, Jones et al. (1987) indicated that plasmid-free segregants of ColE1-derived pMB9- and pBR322-harboring *E. coli* cells arose after approximately 30 generations under both glucose and phosphate limitations in chemostat without any selective pressure, such as antibiotics. On the other hand, when *E. coli* strains harboring the plasmids RP1, pDS4101, or pDS1109 were maintained in such chemostat, plasmid-free segregants were not detected even after 120 generations of nutrient-limited growth.

Similar instability phenomena of plasmids was observed by Noack et al. (1981). Their experiments showed that the plasmid vector pBR325 was unstable in chemostat in the absence of antibiotic selection under both glucose and ammonium chloride limitations, while a similar plasmid vector pBR322 was stably maintained.

Mizutani et al. (1985) investigated the effects of medium compounds and selective pressure upon the stability of a recombinant plasmid pHI301 carrying cloned α-amylase gene in chemostat culture. When grown in L broth containing 1 g of bacto-tryptone, 0.5 g of yeast extract, 5 g of NaCl, 1.23 g of $MgSO_4$ with antibiotic (ampicillin) selection, the host strain, *E. coli* HB101, maintained the plasmid pHI301 stably. However, when the GEM was grown in a synthetic medium supplemented with glucose as a sole carbon source, the plasmid-free segregants appeared even with antibiotic selection. Further, when soluble starch was used as a sole carbon source, which might be utilized only by the pHI301-harboring cells, the plasmid was reasonably stable.

The appearance of plasmid-free cells, or the loss of the complete plasmid, is called *segregational instability,* and is one of the two major types of plasmid instability. As explained earlier, the appearance of plasmid-free segregants may depend on the genetic background of host-plasmid systems, composition of the medium used for the cultivations, nutrient limitations, selective pressure on plasmids, etc. Though not described above, temperature and the dilution rate in the chemostat culture also had important effects on the behavior of the plasmid-free population (Mizutani et al., 1985; Koizumi et al., 1985).

The other major type of plasmid instability is *structural instability,* which refers to the change in plasmid structure due to insertion, deletion, or rearrangement of DNA, often resulting in the loss of expression of the desired cloned gene.

In the study concerning the stability of the low copy number naturally occurring TOL plasmid pWWO in *P. putida* mt-2 under nonselective conditions in continuous culture, the structural instability of the plasmid was described (Duetz and Andel, 1991). *P. putida* mt-2 harboring the plasmid pWWO, which is considered stable and a nonproducer of plasmid-free segregants, was grown in chemostat under succinate, sulphate, ammonium, or phosphate limitation at different dilution rates. It was commonly observed that the fraction of mutant cells lacking the plasmid-coded enzymes for the degradation of toluene and xylene emerged. Genetic analysis revealed that all the mutants isolated harbored partially deleted plasmids deficient in the toluene and xylene catabolic genes coded on pWWO. The mutant cells increased as a fraction of the total population, because cultivation was prolonged under certain conditions due to the growth-rate advantage.

Mizutani et al. (1985) also reported the appearance of the mutant harboring a partially deleted plasmid in the continuous culture of *E. coli* HB101 harboring a recombinant plasmid pHI301 containing α-amylase gene. Most of the mutants showed plasmid-coded ampicillin resistance but lacked α-amylase activity. Analysis of the plasmid isolated from one such mutant, pHI301-L, revealed that an approximately 2-kb DNA fragment around a *Bam*HI site of the cloned α-amylase gene was deleted.

Under long-term starvation-survival conditions (for more than 600 days in sterile well water), GEMs have often lost the ability to produce a functional plasmid product in spite of the fact that they maintained plasmids (Caldwell, 1989). The same phenomenon occurs in chemostat. Investigators suggested that the reason for this instability of plasmid expression may be that plasmids were altered in the host strains by partial or complete deletions.

The structural instability of plasmids is considered to occur with a certain frequency in any combination of plasmids and host strains under various environmental conditions. The behavior of the formed mutants may depend on their growth characteristics under given conditions.

6.1.2 Instability Mechanisms: Plasmid Partitioning

The stability of a recombinant plasmid is considered to depend on: (1) the frequency of the initial segregational or structural instability event and (2) the population dynamics of parental cells (cells harboring complete plasmids) and modified cells (plasmid-free cells or cells harboring structurally changed plasmids).

Although the mechanisms of the structural instability of plasmids have not been well-elucidated—because of the enormously wide variety of the initial events such as insertion, deletion, or rearrangement of DNA—several common mechanisms or factors concerning the initial segregational instability phenomena have been found and studied.

Most plasmids are lost from the host cell at high or low frequencies due to improper partitioning of the plasmids during cell division. When two daughter cells are generated from a plasmid-harboring mother cell, if no plasmids are given to one of the daughter cells, a plasmid-free segregant should appear in the culture. The probability of formation of a plasmid-free daughter cell is considered to depend on the plasmid copy number in the mother cell, and the distribution mechanism of plasmids to daughter cells at cell division.

When it is assumed that plasmids are partitioned to daughter cells at random, the probability of plasmid loss or of formation of a plasmid-free daughter cell from a plasmid-harboring mother cell, p, is given by the following relation, taking for granted that the probability of a plasmid being distributed into each daughter cell at cell division is equal to 2^{-1}.

$$p = 2_N C_0 (2^{-1})^N (2^{-1})^0 = 2^{1-N} \tag{6.1}$$

where N indicates the copy number of plasmids in the mother cell at cell division.

From Equation (6.1), it is apparent that the frequency of the formation of a plasmid-free segregant is very high when the plasmids are maintained at relatively low copy number, while the frequency may be extremely low when they are present at a high copy number per cell. For example, if the mother cell contains the plasmid at 1, 5, 10, or 20 copies per cell, the possibilities are estimated at 1.0 (a plasmid-free cell will always appear), 6.3×10^{-2}, 2.0×10^{-3}, or 1.9×10^{-6}, respectively.

It is generally considered that high copy number plasmids, which are popular cloning vectors, are partitioned randomly to daughter cells at cell division, formation of the plasmid-free segregants of such plasmid-harboring GEMs may be expressed by Equation (6.1). Consequently the probability may be estimated at very low values. However, the apparent multimerzation of recombinant plasmids derived from naturally occurring high copy number plasmids is considered to enhance the possibility of segregation – multimerzation reduces the number of independently segregating plasmid units (Summers and Sherratt, 1984).

On the other hand, genetic functions preventing the formation of plasmid-free segregants at cell division, ensuring that both daughter cells contain at least one copy of the plasmid, have been found coded on many intermediate or low copy number plasmids (Roberts et al., 1990). Such plasmid-stabilizing mechanisms operate either singly or in combination to enhance plasmid stability during cell division. If the locus of the plasmid specifying the stabilization mechanisms is damaged by mutation, plasmid inheritance becomes unstable and frequent segregational instability occurs at high frequencies. Factors affecting stabilization include:

- Increasing the frequency of plasmid replication initiation may increase the total number of copies available for random distribution, and thus stabilize the plasmid.
- The intramolecular resolution of plasmid multimers, known as multimer resolution, should stabilize plasmids by increasing the number of discrete copies available for segregation.
- Selective killing of plasmid-free segregants contributes to stable plasmid maintenance in the culture population. Plasmid-free cells generated during cell division are selectively killed by a host-lethal product coded on the plasmid which also codes for an inhibitor specific to this host-lethal product.
- Stabilizing function-encoding loci – *partitioning loci* – are thought to ensure the physical separation of plasmid copies to the daughter cells via membrane association, similar to the processes found in bacterial chromosome segregation.

Thus, the initial segregational instability of plasmids is thought to be considerably influenced by genetic inheritance of plasmids. Furthermore, since the copy number of plasmids per cell is influenced by environmental factors such as nutritional conditions, cell concentration, agitation, temperature, different modes of reactor operation, etc. in general, they may also affect the segregational instability of plasmids, especially if there are no specific stabilization mechanisms.

6.1.3 Instability Mechanisms: Plasmid Effects on Growth Rate

Once plasmid-free segregants appear in the culture of plasmid-harboring organisms, the culture becomes a mixed population consisting of plasmid-harboring cells and plasmid-free cells. It is possible that the specific growth rates of plasmid-harboring cells and plasmid-free cells differ significantly. Therefore, the plasmid instability problem should involve the population dynamics of both cells—that may be determined by the difference in the specific growth rate between plasmid-harboring and plasmid-free cells under certain conditions.

In the absence of any selective pressure for the plasmid (or if the plasmid does not encode any essential or advantageous gene), it is generally considered that the plasmid-free cells can attain a higher specific growth rate than plasmid-harboring cells. Consequently, under such cultivation conditions, plasmid-free segregants will out-compete the plasmid-harboring fraction and come to predominance. This growth-rate advantage of plasmid-free segregants, or the attenuation of the growth rate of plasmid-harboring cells has been well-investigated, and it has been suggested that it is a consequence of the use of cellular resources on plasmid-related functions in the plasmid-harboring cells, such as plasmid replication, DNA transcription, and plasmid-encoded mRNA translation (plasmid burden).

The magnitude of the attenuation of growth rate caused by plasmid burden depends upon cellular plasmid content (expressed by mg of plasmid DNA per g of cell, etc.), which is determined by its copy number and size in principle and plasmid gene expression.

Seo and Baily (1985) investigated the influence of plasmid content on host growth and cloned gene product synthesis using *E. coli* HB101 host and plasmid RSF1050, and four other closely related copy number mutant plasmids—all enabling constitutive expression of the enzyme β-lactamase in order to maintain as much similarity as possible in all genetic and metabolic factors except plasmid copy number. Experiments were designed to characterize growth properties of different recombinant strains carrying these different plasmids in various media. Specific growth rates of all recombinant strains with plasmids were lower than for the plasmid-free host, and the effect of plasmid burden was confirmed. Specific growth rate decreased as the plasmid copy number—cellular plasmid content, in this case—increased. The specific growth rates of recombinants harboring plasmids pMD247, pMD246, RSF1050, pMD248, and pFH118—of which copy numbers in *E. coli* HB101 were 12, 24, 60, 122, and 148 per cell, respectively—were 0.92, 0.91, 0.87, 0.82, and 0.77 of that of the plasmid-free host strain in LB medium. The trend of growth rate variation with respect to type of plasmid was the same in all media (in addition to LB broth, M9 minimal medium containing glucose as a sole carbon source, and M9 medium plus casamino acids [M9C] were used), although the growth rate attenuation tended to be greatest in M9 medium, and smallest in M9C medium. Similar trends were observed when the plasmid content was altered by using plasmids of varying sizes.

The influence of plasmid-coded protein expression on growth rate attenuation was investigated by Bentley et al. (Bentley et al., 1990). The host/plasmid system used was *E. coli* RR1 and pBR329 (*Ap, Cm, Tc*). By introducing chloramphenicol at different levels into the culture, the expression or activity of the resistance protein was enhanced, as was the reduction in growth rate brought about by the expression of chloramphenicol-acetyl-transferase and β-lactamase. Results indicated a nearly linear decrease in growth rate with increasing foreign protein content. Also, the change in growth rate due to foreign protein depended on the growth rate of the cells. Interestingly, the observed linear relationship between growth rate and expression of foreign protein was media independent.

From these experimental observations, the effect of the change in plasmid burden (burden factor), a, with varying plasmid content in a cell, c, can be quantified through the empirical expression (Lee et al., 1985):

$$a = (1 - (c/cm))^m \qquad (6.2)$$

where the value of cm (mg of plasmid DNA/g-cell) and the exponent, m, are, in general, likely functions of several factors, including growth conditions and genetic properties of the host and plasmid.

It has been considered that not only plasmid-free segregants but also certain structurally modified plasmid-harboring mutants had growth-rate advantage over the original strain, which harbored a complete plasmid.

For example, Duetz et al. investigated the stability of *P. putida* mt-2 harboring the TOL plasmid pWWO during unlimited growth on benzoate, and observed that the gradual decrease in population harboring the complete TOL plasmid was caused predominantly by a growth-rate advantage of spontaneous mutants harboring a partially deleted plasmid (Duetz et al., 1991). The growth-rate difference between the mutants and original *P. putida* strain mt-2 (pWWO) was 0.11 hour^{-1} at dilution rates from 0.68–0.79 hour^{-1}. It was suggested that the growth-rate disadvantage of the parental strain was caused by inhibitory effects of an intermediate in the degradation of benzoate via plasmid-coded *meta*-cleavage pathway.

6.1.4 Kinetics

Imanaka and Aiba (1981) expressed a mixed-culture system with plasmid-harboring cells and plasmid-free cells by the following three processes for cell division.

$$X^{+} \xrightarrow{(\mu^{+})} (2 - p)X^{+} \qquad (6.3a)$$

$$\longrightarrow pX^{-} \qquad (6.3b)$$

$$X^{-} \xrightarrow{(\mu^{-})} 2X^{-} \qquad (6.3c)$$

where X^{+} and X^{-} indicate the plasmid-harboring parental strain and plasmid-free strain, μ^{+} and μ^{-} are the specific growth rates of the plasmid-harboring cells and plasmid-free cells, respectively, and p is the probability of plasmid loss at cell division.

Accordingly, the dynamics of X^{+} and X^{-} are given by a set of ordinary differential equations.

$$dX^{+}/dt = (1 - p)\mu^{+}X^{+} \qquad (6.4a)$$

$$dX^{-}/dt = \mu^{-}X^{-} + p\mu^{+}X^{-} \qquad (6.4b)$$

where t is time.

If the plasmid content of X^{+} and gene expression are maintained at a certain constant level, the relationship between μ^{+} and μ^{-} is considered as follows:

$$\mu^{-} = \alpha\mu^{+} \qquad (6.5)$$

where α is the ratio of the specific growth rate of plasmid-free cells to that of plasmid-harboring cells, indicating the magnitude of growth advantage of the plasmid-free population.

Therefore, the growth dynamics of X^{+} and X^{-} can be expressed by two parameters, p and α, with Equations (6.4a), (6.4b), and (6.5).

6.2 SCREENING OF STABLE GEMs FOR WASTEWATER TREATMENT

6.2.1 Plasmid Stability Test

As mentioned above, the stability of recombinant plasmids encoding desired genes in the host strains depends on the genetic properties of host/plasmid systems and the environmental conditions for cultivation. Since it is impossible or very difficult to keep any selective pressure for GEMs or recombinant plasmids consistently in wastewater treatment processes—because of both qualitative and quantitative fluctuation of influent, and uncontrollable environmental factors such as temperature—it is necessary to select host/plasmid systems having high genetic stability in the absence of selective pressure. Therefore, we investigated the stability of recombinant plasmids containing catechol-degrading *pheB* gene in a variety of host strains by means of serially transferred culture using LB broth without any antibiotic selection.

GEMs used in the screening were *E. coli* and *P. putida* strains harboring recombinant plasmids pBH100 or pBH500. The pBH100 was constructed by cloning a 5.65-kb DNA fragment containing the *pheB* gene from the chromosome of *P. putida* BH in pUC19, and introduced only into *E. coli* strains. The pBH500 was constructed by inserting the cloned DNA fragment into a broad host range vector pKT230. The details of both plasmids were described in Section 5.1. The pBH100 was transformed into *E. coli* JM103 and *E. coli* C600, while the pBH500 into *E. coli* C600, *P. putida* KT2440, *P. putida* PpG1064, and *P. putida* BH-1.

These GEMs were precultured in 10 mL of LB broth supplemented with ampicillin for pBH100 or streptomycin for pBH500, then 0.1 mL of each culture was inoculated into 10 mL of antibiotic-free LB broth. After 20–26 hrs of incubation at 30°C on a reciprocal shaker at 90 rpm (37 mm strokes), 0.1 mL of culture broth was transferred into a new 10 mL of LB broth. This transfer was repeated serially, and the culture broth was sampled and spread onto LB agar plates periodically to investigate the phenotypes of the culture population. Approximately 150–300 colonies from each plate were tested for resistance to the corresponding antibiotic (Ap or Sm) as a plasmid marker by transferring with tooth picks to antibiotic-containing LB agar plate. Positive colonies grew on the selective medium. The *pheB*-coded C23O activity, another plasmid marker, was also investigated by spraying catechol solution on the colonies—positive colonies turned yellowish due to the accumulation of 2-HMS.

Results of such phenotypic analysis in the stability tests are summarized in Table 6.1. According to the above-mentioned phenotypic analysis, in addition to the original parental population which possessed both antibiotic resistance and constitutive expression of *pheB,* (Ap^r, $C23O^+$) or (Sm^r, $C23O^+$) for pBH100- or pBH500-harboring GEMs respectively, the subpopulation which showed the phenotype of antibiotic-sensitive and C23O-negative, (Ap^s, $C23O^-$) or (Sm^s, $C23O^-$), was detected for all GEMs tested. Several colonies which lacked the expression of both plasmid markers (Ap^s, $C23O^-$) or (Sm^s, $C23O^-$) were picked up randomly and examined for

TABLE 6.1. Phenotypic analysis in the plasmid stability test.

Test Strain	Phenotype			
	(Ap^r/Sm^r, C230$^+$)	(Ap^r/Sm^r, C230$^-$)	(Ap^s/Sm^s, C230$^+$)	(Ap^s/Sm^s, C230$^-$)
E. coli JM103 (pBH100)	466	0	0	234
E. coli C600 (pBH100)	149	0	0	271
E. coli C600 (pBH500)	1,455	24	0	1,458
P. putida KT2440 (pBH500)	1,180	0	0	606
P. putida PpG1064 (pBH500)	1,159	0	0	1,674
P. putida BH-1 (pBH500)	2,024	0	0	223

*Expressed as no. of colonies. Phenotypes are shown in the text.

the possession of plasmid(s) by the extraction method of Birnboim and Doly (1979). Agarose gel electrophoresis analysis of the crude extract from these (Ap^s, C230$^-$) or (Sm^s, C230$^-$) strains showed that all of them were plasmid-free segregants—no plasmid bands were detected. Thus, the segregational instability of the recombinant plasmid was observed in each GEM.

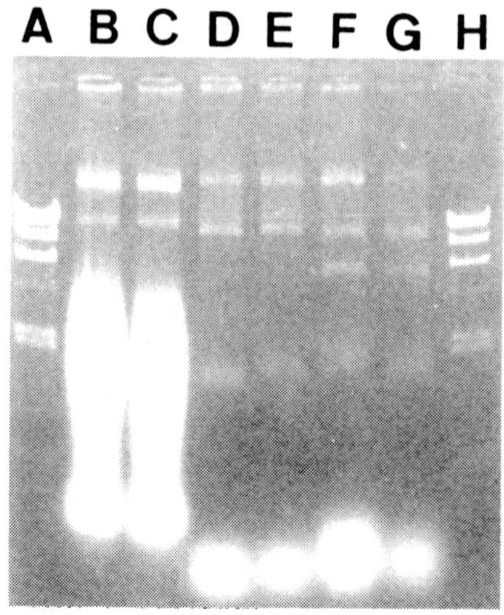

FIGURE 6.1. Plasmid analysis of a (Sm^r, C230$^-$) strain: lane A, H, *Hin*dIII-digested λDNA; lane B, crude plasmid extract from a (Sm^r, C230$^-$) strain; lane C, crude plasmid extract from a (Sm^r, C230$^+$) strain; lane D, *Hin*dIII-digested (Sm^r, C230$^-$) plasmid; lane E, *Hin*dIII-digested (Sm^r, C230$^+$) plasmid; lane F, *Hin*dIII-*Bam*HI-digested (Sm^r, C230$^-$) plasmid; lane G, *Hin*dIII-*Bam*HI-digested (Sm^r, C230$^+$) plasmid.

On the other hand, in the stability test of *E. coli* C600 (pBH500), another subpopulation which showed antibiotic-resistance but lacked C230 activity, (Sm^r, $C230^-$), was detected at certain low frequencies during cultivation. Some (Sm^r, $C230^-$) strains were also selected randomly and examined for the presence of plasmid by plasmid extraction and agarose gel electrophoresis. The results of the electrophoresis of the plasmid extracted from one (Sm^r, $C230^-$) strain are shown in Figure 6.1. As shown in the figure, an approximate 17.5-kb plasmid band that was similar or identical to the pBH500 in size was observed when digested with *Hin*dIII, and this observation was common for all examined (Sm^r, $C230^-$) strains. Although one such (Sm^r, $C230^-$) strain was designated *E. coli* C600-W304, and its physiological characteristics and the physical structure of its plasmid (restriction endonuclease analysis and DNA-DNA hybridization by using the labeled 5.65-kb *pheB*-containing DNA fragment cloned from *P. putida* BH as a probe) were investigated further, no differences between the parental GEM *E. coli* C600 and *E. coli* C600-W304 were discovered except for the C230 activity. The C230 activity of *E. coli* C600 (pBH500) was 73.0 Units/g-protein, while that of *E. coli* C600-W304 was 0.3 Units/g-protein. That of the plasmid-free host was below 0.1 Units/g-protein. Thus, *E. coli* C600-W304 was considered a mutant in which the expression of the *pheB* gene coded on pBH500 was extremely low. When plasmid extracted from *E. coli* C600-W304 by the Birnboim and Doly method was used to transform *E. coli* C600, all the transformants showed the phenotype (Sm^r, $C230^-$). The low expression of *pheB* (low C230 activity) should be due to the structural mutation on the recombinant plasmid pBH500. However, this mutation has not yet been clarified fully. Thus, in addition to segregational instability, structural instability was observed for *E. coli* C600 (pBH500).

6.2.2 Stable GEMs

Figure 6.2 and Figure 6.3 show the time profiles for the fraction of plasmid-harboring population in the serially transferred cultivation of pBH100- and pBH500-harboring GEMs, respectively. For *E. coli* C600, the fraction of the phenotype (Sm^r, $C230^-$), such as *E. coli* C600-W304, is also shown in Figure 6.3(a). In these figures, time is expressed by generation defined as

$$2^n = OD_t/OD_0 \tag{6.6}$$

where n is generation of culture at time t, OD_t indicates the density of grown cells measured as OD_{600} at time t, and OD_0 indicates the density of the inoculated cells at time zero.

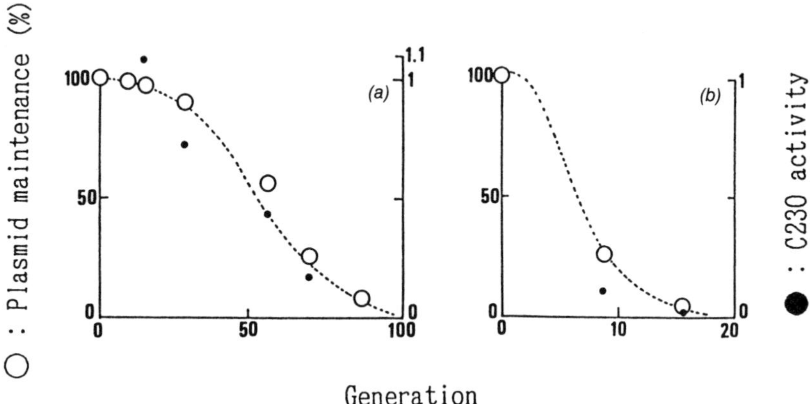

FIGURE 6.2. Plasmid stability of pBH100-harboring GEMs: (a) stability test of JM103 (pBH100); (b) stability test of C600 (pBH100). Plasmid maintenance (○) is expressed by the fraction of plasmid-harboring cells. C230 activity (●) is expressed by the ratio to initial value.

As shown in Figure 6.2 and Figure 6.3, the population harboring the complete recombinant plasmid pBH100 (Ap^r, $C230^+$) or pBH500 (Sm^r, $C230^+$) decreased due to the increase of the plasmid-free segregants (Ap^s, $C230^-$) or (Sm^s, $C230^-$), as time progressed in the stability tests of all GEMs. Ultimately, the culture of each GEM was dominated or conquered by the plasmid-free segregants. On the other hand, the subpopulation of plasmid-harboring mutants which showed the phenotype (Sm^r, $C230^-$) was maintained at low percentages during the stability test of *E. coli* C600 (pBH500). From these experimental results it was confirmed that segregational instability is a very important problem—in comparison to structural instability—in the cultivation of plasmid-harboring GEMs without any selective pressure.

Although it was commonly observed that the culture went through a stable lag phase followed by a rapid increase in the plasmid-free segregants in all tests, the degrees of the rates of increase of plasmid-free segregants, or the decrease of plasmid-harboring cells, were different. In other words, the stability of recombinant gene *pheB* depended on the combination of the host strains and plasmids (pBH100 or pBH500).

For example, approximately 75% of the total cells were plasmid-free segregants after nine generations in the cultivation of *E. coli* C600 (pBH100), whereas the population of plasmid-harboring GEM accounted for more than 95% after the same number of generations in the culture of *E. coli* JM103 (pBH100), indicating that the recombinant plasmid, pBH100,

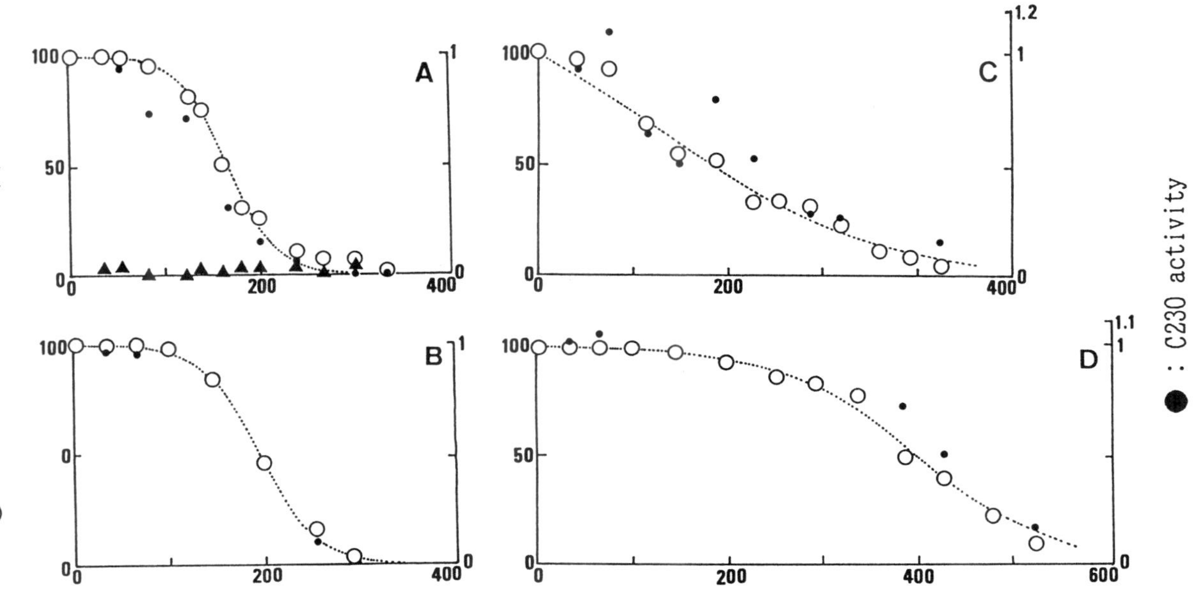

FIGURE 6.3. Plasmid stability of pBH500-harboring GEMs: (a) stability test of C600 (pBH500); (b) stability test of KT2440 (pBH500); (c) stability test of PpG1064 (pBH500); (d) stability test of BH-1 (pBH500). Plasmid maintenance (○) is expressed by the fraction of plasmid-harboring cells. Fraction of the mutant population (Sm^r, $C230^-$) is also shown in (c) by ▲. C230 activity (●) is expressed by the ratio to initial value. (Reprinted from *Water Research, Vol. 25,* Fujita, Ike and Hashimoto, "Feasibility of Wastewater Treatment Using Genetically Engineered Microorganisms," p. 983, 1991, with kind permission from Pergamon Press Ltd., Headington Hill Hall, Oxford OX3 0BW, UK.

was maintained much more stably in *E. coli* JM103 than in *E. coli* C600. The pBH500 was even more stable in the host *E. coli* C600. After about 100 generations, the fractions of plasmid-free segregants of *E. coli* C600 (pBH100) and *E. coli* C600 (pBH500) were approximately 95% and 5%, respectively.

In general, it is considered that as the expression of cloned gene products coded on the recombinant plasmid increases, the rate of growth of plasmid-free segregants also increases because of the metabolic burden. From this viewpoint, at the higher level a GEM expresses the cloned gene, the less stability the recombinant plasmid has in the host strain. However, the experimental results obtained here did not always agree with this theory (see Table 5.2, Figure 6.2, and Figure 6.3). Among the four pBH500-harboring GEMs, the expression of *pheB* gene (C230) activity was *E. coli* C600 (pBH500) > *P. putida* PpG1064 (pBH500) > *P. putida* BH-1 (pBH500) ≥ *P. putida* KT2440 (pBH500), while the rate of appearance or increase of plasmid-free segregants was *E. coli* C600 (pBH500) ≥ *P. putida* KT2440 (pBH500) > *P. putida* PpG1064 (pBH500) > *P. putida* BH-1 (pBH500). These results suggested that we can choose the host/vector systems suitable for wastewater treatment which have both high expression of cloned genes and high plasmid stability. The most suitable host strain for pBH500 among the four strains used in this study may be *P. putida* PpG1064, as it showed both relatively high gene expression and plasmid stability. *P. putida* BH-1 may be another suitable host because of its extremely high stability; after approximately 300 generations, more than 80% of the cells maintained the recombinant plasmid pBH500. One putative reason for this extremely high stability is considered to be that the *pheB*-encoding DNA fragment contained in the recombinant plasmid pBH500 is derived from the chromosome of the host strain, *P. putida* BH-1 (or *P. putida* BH), consequently the burden of this host strain for the expression of *pheB* may be less than the other host strains.

6.2.3 Model Analysis

As mentioned in Section 6.1, it is generally considered that the stability of plasmid depends on two major parameters, the probability of plasmid loss due to segregation during cell division and the difference in the specific growth rate between plasmid-harboring cells (GEMs) and plasmid-free segregants, if other mutant cells, such as *E. coli* C600-W304, do not appear in the culture at a considerably high level.

Based on the kinetics of plasmid-harboring cells and plasmid-free cells described in Section 6.1.4 [the processes (6.3a), (6.3b), and (6.3c) or the Equations (6.4a) and (6.4b)], Imanaka and Aiba (1981) expressed the time profile of the fraction of plasmid-harboring cells in culture maintained in the exponential growth phase by the following equation:

$$Fn = (1 - \alpha - p)/\{1 - \alpha - p \cdot 2^{n(a+p-1)}\} \qquad (6.7)$$

where *Fn* is the fraction of plasmid-harboring cells after *n* generations, *p* is the probability of generation of a plasmid-free segregant, and α is the ratio of the specific growth rate of plasmid-free cells to that of plasmid-harboring cells defined in Equation (6.5).

In order to analyze and quantify the two parameters, α and *p*, which contribute to the plasmid maintenance curves or the time profiles of plasmid-harboring populations, the experimental data obtained in the serially transferred culture that might satisfy the assumptions of Equation (6.7) as a whole were fitted to this model equation, and values of α and *p* were estimated by minimizing the least squares of differences between the data and the model curves. Estimated values of the two parameters are listed in Table 6.2, and theoretical curves obtained from these values are shown as the solid lines together with the data points in Figure 6.2 and Figure 6.3. As shown in the figures, the model curves of Equation (6.7) provided excellent fits to the experimental data, with the only exception the case of *P. putida* PpG1064 (pBH500).

Estimated values of *p*, the probability of plasmid loss, were low, and affected little by the host strain. On the other hand, the *p* value depended upon the plasmids—10^{-4} and 10^{-5} orders for pBH100- and pBH500-harboring GEMs, respectively, except for *P. putida* PpG1064 (pBH500), suggesting that pBH100 constructed with the high copy number plasmid vector pUC19 tends to be lost more often from the host strains at cell division than the pBH500 constructed with medium copy number plasmid pKT230. Interest-

TABLE 6.2. Parameter estimates for plasmid stability of GEMs.

Strain	α	*p*	No. of Data	Σ *
E. coli JM103 (pBH100)	1.13	0.0009	7	0.0039
E. coli C600 (pBH100)	2.96	0.0004	3	0.0008
E. coli C600 (pBH500)	1.06	0.00006	13	0.0249
P. putida PpG1064 (pBH500)	1.01	0.003	13	0.0336
P. putida KT2440 (pBH500)	1.05	0.00006	8	0.0022
P. putida BH-1 (pBH500)	1.02	0.00008	13	0.0163

*Σ : residual sum of squares.

Reprinted from *Water Research, Vol. 25,* Fujita, Ike and Hashimoto, "Feasibility of Wastewater Treatment Using Genetically Engineered Microorganisms," p. 983, 1991, with kind permission from Pergamon Press Ltd., Headington Hill Hall, Oxford OX3 0BW, UK.

ingly, neither pUC19 nor pKT230 have any stabilizing mechanisms for plasmid maintenance. However, the p value seemed to have little or no relation to the stability of the plasmids in the experiments here.

On the other hand, the α value for each GEM was estimated at more than 1.0 as anticipated [see Section 6.1.4, Equation (6.5)], and it was shown that the larger the estimated value of α, the faster plasmid stability decreased. Therefore, plasmid stability is considerably affected by the magnitude of growth advantage of the plasmid-free population (α). Thus, in order to construct a GEM with high genetic stability, it may be important to select a host strain whose specific growth rate is little affected or decreased by plasmid maintenance.

6.2.4 Conclusions and Considerations

The genetic stability of various GEMs carrying the *pheB* gene coded on the recombinant plasmid pBH100 or pBH500 was investigated by means of serially transferred culture in LB broth which has no selective pressure on the GEMs. Although both segregational and structural instability of the recombinant plasmids were observed through the experiments, the former was considered more important than the latter, because the plasmid-free segregants derived from all GEMs tested, while the mutants—which possessed maybe a small amount of modified plasmids and lacked the expression of the recombinant gene *pheB*—appeared only in the cultivation of *E. coli* C600 (pBH500) among six GEMs used in this study.

Although it was commonly observed that the culture went through a stable lag phase followed by a rapid increase in the plasmid-free segregants in all tests, the rates of increase of plasmid-free segregants were different. These results suggested that the stability of recombinant plasmid depends upon the host/vector system. Experimental results also suggested that we can choose host/vector systems suitable for wastewater treatment which have both high expression of cloned genes and high plasmid stability. One such suitable GEM selected in this study was *P. putida* BH-1 (pBH500), which had extremely high plasmid stability—after approximately 300 generations, more than 80% of the culture population maintained the plasmid. One putative reason for this extremely high stability is that the *pheB*-encoding DNA fragment cloned into the recombinant plasmid pBH500 is derived from the chromosome of the host strain, consequently the burden of this host strain for the expression of *pheB* may be relatively light. Therefore, reintroduction of the cloned gene into the parental host strain (see Section 5.3) may generate GEMs which have high genetic stability.

Further, the model analysis of the experimental data showed that plasmid stability is affected considerably by the magnitude of growth advantage of the plasmid-free population. Therefore, to construct a GEM with high genetic stability, it is considered important to select the host strain whose specific growth rate is little affected or decreased by plasmid maintenance.

6.3 REFERENCES

Bentley, W. E., N. Mirjalili, D. C. Anderson, R. H. Davis and D. S. Kompala. 1990. "Plasmid-Encoded Protein: The Principal Factor in the "Metabolic Burden" Associated with Recombinant Bacteria," *Biotechnol. Bioeng.*, 35:668–681.

Birnboim, H. C. and J. Doly. 1979. "A Rapid Alkaline Extraction Procedure for Screening Recombinant Plasmid DNA," *Nucleic Acid Res.*, 7:1513–1323.

Caldwell, B. A., C. Ye, R. P. Griffiths, C. L. Moyer and R. Y. Morita. 1989. "Plasmid Expression and Maintenance during Long-Term Starvation–Survival of Bacteria in Well Water," *Appl. Environ. Microbiol.*, 55:1860–1864.

Duetz, W. A., M. K. Winson, J. G. Van Andel and P. A. Williams. 1991. "Mathematical Analysis of Catabolic Function Loss in a Population of *Pseudomonas putida* mt-2 during Non-Limited Growth on Benzoate," *J. Gen. Microbiol.*, 137:1363–1368.

Duetz, W. A. and J. G. Van Andel. 1991. "Stability of TOL Plasmid pWWO in *Pseudomonas putida* mt-2 under Non-Selective Conditions in Continuous Culture," *J. Gen. Microbiol.*, 137:1369–1374.

Imanaka, T. and S. Aiba. 1981. "A Perspective on the Application of Genetic Engineering: Stability of Recombinant Plasmid," *Ann. N.Y. Acad. Sci.*, 369:1–14.

Jain, R. K., G. S. Sayler, J. T. Wilson, L. Houston and D. Pacia. 1987. "Maintenance and Stability of Introduced Genotypes in Groundwater Aquifer Material," *Appl. Environ. Microbiol.*, 53:996–1002.

Jones, I. M., S. B. Primrose, A. Robinson and D. C. Elwood. 1980. "Maintenance of Some ColE1-Type Plasmids in Chemostat Culture," *Molec. Gen. Genet.*, 180:579–584.

Koizumi, J., Y. Monden and S. Aiba. 1985. "Effects of Temperature and Dilution Rate on the Copy Number of Recombinant Plasmid in Continuous Culture of *Bacillus stearothermophilus* (pLP11)," *Biotechnol. Bioeng.*, 27:721–728.

Lee, S. B., A. Seressiotis and J. E. Bailey. 1985. "A Kinetic Model for Product Formation in Unstable Recombinant Populations," *Biotechnol. Bioeng.*, 27:1699–1709.

Mizutani, S., S. Fukuzono, N. Tsukaguchi, S. Ueda and T. Kobayashi. 1985. "Stability of a Recombinant Plasmid Containing α-Amylase Gene in Chemostat," *J. Chem. Eng. Jap.*, 18:220–224.

Noack, D. M., M. Roth, R. Geuther, G. Muller, K. Undisz, C. Hoffmeier and S. Gaspar. 1981. "Maintenance and Genetic Stability of Vector Plasmids pBR322 and pBR325 in *Esherichia coli* K12 Strains Grown in a Chemostat," *Mol. Gen. Genet.*, 184:121–124.

Roberts, R. C., R. Burioni and D. R. Helinski. 1990. "Genetic Characterization of the Stabilizing Function of a Region of Broad-Host-Range Plasmid RK2," *J. Bacteriol.*, 172:6204–6216.

Seo, J.-H. and J. E. Bailey. 1985. "Effects of Recombinant Plasmid Content on Growth Properties and Cloned Gene Product Formation in *Escherichia coli*," *Biotechnol. Bioeng.*, 27:1668–1674.

Summers, D. K. and D. J. Sherratt. 1984. "Multimerization of High Copy Number Plasmids

Causes Instability: ColEl Encodes a Determinant Essential for Plasmid Monomerzation and Stability," *Cell.*, 36:1097–1103.

Weber, A. E. and K.-Y. San. 1990. "Population Dynamics of a Recombinant Culture in a Chemostat under Prolonged Cultivation," *Biotechnol. Bioeng.*, 36:727–736.

Williams, P. A., S. D. Taylor and L. E. Gibb. 1988. "Loss of the Toleune-Xylene Catabolic Genes of TOL Plasmid pWWO during Growth of *Pseudomonas putida* on Benzoate is Due to a Selective Growth Advantage of 'Cured' Segregants," *J. Gen. Microbiol.*, 134:2039–2048.

CHAPTER 7

Ecological Stability of Genetically Engineered Microorganisms in Wastewater Treatment Processes

Another stability problem of GEMs in their application to wastewater treatment processes is ecological stability which occurs only when GEMs are introduced into mixed microbial communities. *Ecological stability* refers to the length of time GEMs can survive and express their useful activities in the mixed community of a wastewater treatment process. It also refers to the population size GEMs can maintain there. Even if a GEM with a high xenobiotic compound degradation rate is constructed, it may soon disappear from the process or decrease dramatically if it has low ecological stability, and the treatment efficiency will thus be little improved. The survival of introduced bacterial strains, including GEMs, in the mixed flora of natural environments or microcosms has been dealt with in a considerable number of studies that assumed intentional or accidental release. However, there have been only a few such studies on the survival of a specific bacterial strain or a GEM in wastewater treatment processes. One conclusion in these studies investigating the survival of natural or genetically engineered bacteria capable of degrading 3-chlorobenzoate in a laboratory-scale activated sludge unit was that it is important to choose a bacterial strain which is well-adapted to the environmental conditions for the successful use of microbial inoculation (McClure et al., 1991; 1989). To date, little is known about the survival of GEMs introduced into wastewater treatment processes.

This chapter, composed of three sections, will deal with the ecological stability of GEMs in wastewater treatment. The first two sections describe the survival of GEMs in natural environments from references, and in activated sludge processes from our experimental researches. The last section suggests ways to improve the ecological stability or survival of GEMs in activated sludge processes.

7.1 SURVIVAL OF GEMs IN NATURAL ENVIRONMENTS

Recently, numerous researches have been performed on the survival of GEMs and manmade DNA in natural environments, such as soil, sediments, and pond and lake water. In general, factors affecting the survival of GEMs and recombinant DNA are considered to be divided into biotic and abiotic components. Biotic factors include gene transfer processes, stability of recombinant DNA molecules, growth and decay property of the GEMs, microbial interactions such as competition with indigenous bacteria and predation by protozoa, etc. Abiotic factors include temperature, oxygen, pH, light level, water content, salinity, nutrients, boundaries, etc. Because the transfer of recombinant DNA molecules was already detailed in Chapter 3, this section will focus on the survival of GEMs.

The survival of various GEMs in soils, where no specific selective pressure was present, showed a similar pattern in many cases. GEMs or natural bacteria which were introduced into soils at certain high inoculum concentrations initially showed decline in number (measured as *cfu* = viable counts) followed by the stabilization of population densities at constant levels. This fit the survival pattern in soils of *P. putida* EEZ15 carrying a recombinant TOL plasmid pWWO-EB62 (Ramos et al., 1991), *Rhizobium* sp. (Beringer and Bale, 1988), and *P. putida* bearing a recombinant plasmid containing *xylE* gene, pLV1013 or pLV1016 (Macnaughton et al., 1992). However, the presence of a certain selective pressure caused fast growth or abrupt decline of the introduced GEMs, which differs from the above-mentioned basic pattern (Ramos et al., 1991). Moreover, the survival pattern of GEMs was considerably affected by the soil types used in the experiments, probably depending on the differences in the presence of predators and the available nutrients (Mcnaughton et al., 1992).

The typical survival properties of introduced GEMs in aquatic environments was estimated from the experimental results of Amy and Hiatt (1989). The GEM, *E. coli* HB101, harboring a recombinant plasmid, pPSA131, which was constructed by the ligation of the common cloning vector pBR325 with a plasmid fragment from *Alcaligenes* sp. encoding mercury resistance and 2,4-dichlorophenoxyacetic acid degradation, was inoculated into Lake Mead water (Nevada) and salts buffer. Survival of *E. coli* HB101 (pPSA131) remained high in filtered lake water with a 0.22-μm-pore-size membrane and salts buffer (approximately on the order of 10^5 cfu/mL on day 7), but was lower in untreated lake water or water filtered with a 0.8-μm-pore-size membrane (approximately at 10^4 cfu/mL each). This suggested that the competition with indigenous bacteria caused the decline of the GEM, and that the effect of predation by protozoa (which may be removed by filtration with a 0.8-μm-pore-size membrane) was less. Such important effects of the competition with other organisms indigenous to lake water was also described in the paper dealing with the survival of genetically engineered *E. coli* and *P. putida* in lake water microcosms that utilized survival membrane chambers in a flowthrough or static renewal system (Awong et al., 1990). In this study, the survival patterns of the GEMs and their parent host strains (plasmid-free strains) were not much different.

Although many studies showed interesting observations, further studies are considered necessary to obtain data concerning the survival of GEMs in natural environments or mixed microbial communities.

7.2 SURVIVAL OF GEMs IN ACTIVATED SLUDGE PROCESSES

7.2.1 Model Processes

We used two types of model activated sludge treatment processes to evaluate the survival of GEMs in activated sludge microcosms, and to determine the operational parameters which affect it. One was the fill-and-draw system which simulated the fed batch or sequential batch treatment process, and the other was the continuous flow system which was a model for the conventional treatment process.

Activated sludge treatment by the fill-and-draw system was performed by shake culture. The seed activated sludge (100 mL) was cultivated in a 300-mL Erlenmeyer flask at 25°C on a rotary shaker with rotation set at 100 rpm. After approximately 1-day (21–25 hours) cultivation, a definite volume of waste sludge was withdrawn to control the SRT of the system, and the rest was settled for 30 min. The supernatant was then discarded as the effluent, and the sterilized synthetic wastewater composed of meat extract, peptone, urea, and minimal salts, was added to the original volume (100 mL). The supernatants removed were approximately 85–90% of the total volume. Such operations were repeated daily, commonly called the fill-and-draw technique.

The continuous flow system was constructed using a bench-scale model unit. The unit consisted of an aeration tank with a liquid overflow into a sedimentation tank. The aeration tank had a working volume of 5 L, and the sedimentation tank 1 L. The settled sludge was returned to the aeration tank from the bottom of the sedimentation tank at the rate of 4.3 L/day. Synthetic wastewater was fed continuously to the aeration tank at the flow rate of 5 L/day. Aeration was conducted at 0.4 vvm with an air pump and diffuser attached at the bottom of the tank, in which the temperature was maintained at 25°C with a thermostat and heater. Waste sludge was withdrawn from the aeration tank. The amount was calculated according to the SRT considering the loss of sludge which was contained in the effluent.

The seed activated sludge used for both systems had been acclimated to the synthetic wastewater used in the study for more than 2 years by the fill-and-draw technique.

7.2.2 Survival in Fill-and-Draw Process

Two types of GEMs were used in this survival study—*E. coli* C600 (pBH500) and *P. putida* BH (pBH500). The properties of the hosts and the recombinant plasmid used for the construction of both GEMs were described in Chapter 5. The recombinant plasmid, pBH500, gives the host strains resistance to streptomycin and the constitutive expression of *pheB* gene. To enumerate the GEMs, the plate count technique using selective media was used. *E. coli* C600 (pBH500) was enumerated by using deoxycholate agar, which is a selective medium for enteric bacteria, supplemented with 100 mg/L streptomycin, and *P. putida* BH (pBH500) by benzoate minimal salts medium with 50 mg/L streptomycin. On these selective media, no background colonies were detected when the activated sludge samples with no GEMs were plated and incubated at 30°C for 1–2 days. The viable bacterial counts of the activated sludge on the selective media were all below 10^0 cfu/mL, although those counted on CGY agar (25°C, 7 days) were 10^9 to 10^{10} cfu/mL.

The GEMs were grown in LB broth and added to 100 mL of the seed sludge in a 300-mL Erlenmeyer flask, of which MLSS concentration was approximately 2,000 mg/L. Their survival was investigated in the fill-and-draw process. Figure 7.1 shows a typical time course of the survival of *E. coli* C600 (pBH500) and *P. putida* BH(pBH500), which were inoculated into the activated sludge at 3.5×10^9 and 2.3×10^9 cfu/mL, respectively, in the fill-and-draw system.

To evaluate the stability of the recombinant plasmid in the host strains, the activated sludge samples were plated onto selective media without streptomycin, which is a selective marker for the pBH500 plasmid, during the experiments, and the appearance of segregants (plasmid-free host cells) was estimated. The results suggested that the ratio of segregants was approximately 5–10% throughout the experimental period. The phenotypes and morphologies of several colonies formed on the selective media were investigated further, and none of the colonies tested could be distinguished from the introduced GEMs. Therefore, the selective media used seemed to be specific enough to enumerate and identify only GEMs.

As shown in Figure 7.1, the number of both GEMs declined rapidly, after which their population remained relatively stable for a considerable time in

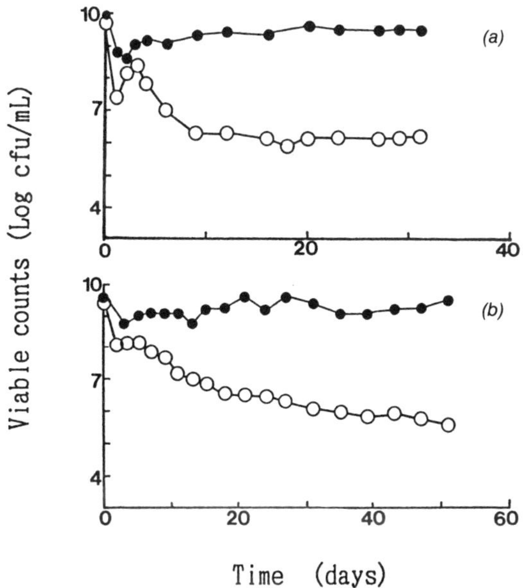

FIGURE 7.1. Survival of GEMs in the fill-and-draw activated sludge process: (a) survival of C600 (pBH500); (b) survival of BH (pBH500); ○, total bacterial count; ●, GEMs.

the fill-and-draw process. For these examples, in which the SRT of the process was maintained at 20 days, the population of *E. coli* C600 (pBH500) decreased from 3.5×10^9 to approximately 1×10^6 cfu/mL in 9 days, and then remained very stable with an occasional small drop up to day 31. On the other hand, *P. putida* BH (pBH500) inoculated at 2.3×10^9 also showed abrupt decline to approximately 5×10^6 cfu in 15 days, and the population was maintained relatively stable with a gradual decline up to day 51. These results suggested that the survival course of GEMs in activated sludge may be divided into two phases—the former period a declining phase and the latter period a stable phase. This pattern occurred in all the survival courses of both *E. coli* C600 (pBH500) and *P. putida* BH (pBH500) obtained in the experiments using the fill-and-draw process under different conditions. However, the degree of gradual decline in the stable phase seemed to be less for *E. coli* C600 (pBH500) than for *P. putida* BH (pBH500). During the experiments, total heterotrophic bacteria, which were counted by using CGY agar (7–10 days incubation at 25°C), were maintained at approximately 10^9 cfu/mL.

The effects of inoculum size on the survival of GEMs in the activated sludge were investigated in the fill-and-draw process. The survival of *E.*

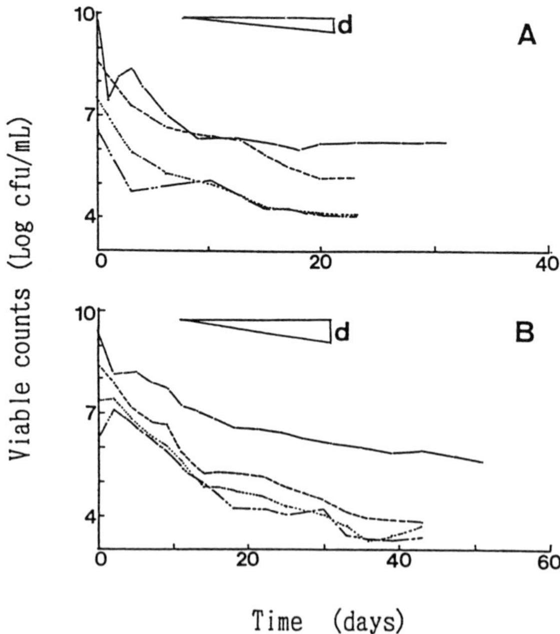

FIGURE 7.2. Effect of inoculum size on the survival of GEMs: (A) effect on the survival of C600 (pBH500); (B) effect on the survival of BH (pBH500); d, theoretically expected decrease due to sludge wastage corresponding to an SRT of 20 days.

coli C600 (pBH500) and *P. putida* BH (pBH500), which were inoculated at different densities, are shown in Figure 7.2. Here, the SRT was set at 20 days in all cases. As shown in the figures, although the inoculum size of the introduced GEMs had little influence upon the rate of decline in the declining phase, in the stable phase the higher the density inoculated, the higher the level of population. In the fill-and-draw process, therefore, the inoculum size seems to affect the survival of GEMs.

The effect of the SRT on the survival of introduced GEMs in the fill-and-draw process was also investigated. *E. coli* C600 (pBH500) was inoculated into four flasks (the fill-and-draw processes) at 3.5×10^8 cfu/mL, and *P. putida* BH (pBH500) at 2.7×10^8 cfu/mL, and the SRTs of flasks were controlled at 3, 5, 10, and 20 days as separate experiments; 33, 20, 10, and 5 mL of the activated sludge (of total 100 mL) per day was withdrawn, respectively. Results are shown in Figure 7.3. As shown in the figures, as the SRT of the process decreased, the population size of GEMs [both *E. coli* C600 (pBH500) and *P. putida* BH (pBH500)] surviving in the stable phase—i.e., ecological stability—increased in spite of the decrease in the MLSS. MLSS in the stable phase was approximately 600–700, 1,000–1,100, 2,000–2,200, and 2,300–2,500 mg/L at SRTs of 3, 5, 10, and 20 days. The increase in ecological stability according to the decrease in SRT was due to the shortening of the period of the declining phase and the gradual decline in the stable phase for *E. coli* C600 (pBH500). For *P. putida* BH (pBH500), on the other hand, it depended upon the lowering of the rate of decline in both the declining and stable phases.

The theoretically expected decline of GEMs due to sludge wastage corresponding to SRTs of 3, 5, 10, and 20 days is described in Figure 7.3 as slopes. This expected decline was calculated both on the assumption that the GEMs did not grow at all in the activated sludge, and that they were contained evenly in the waste sludge. It is apparent, from the comparison of the survival courses and the theoretically expected declines of GEMs in the figures, that the shorter the SRT of the process, the faster the GEMs grew in the activated sludge.

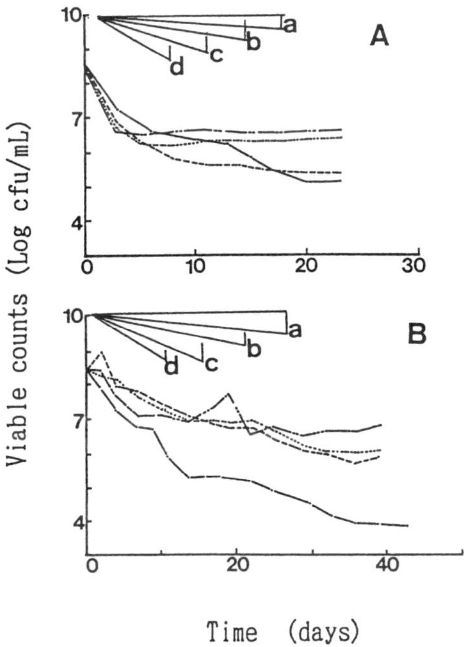

FIGURE 7.3. Effect of SRT control on the survival of GEMs: (A) effect on the survival of C600 (pBH500); (B) effect on the survival of BH (pBH500); SRT was controlled at 20(——), 10(----), 5(------), and 2 days (–·–·–); a, b, c, d, theoretically expected decrease due to sludge wastage corresponding to SRTs of 20, 10, 5, and 3 days, respectively.

7.2.3 Survival in the Continuous Flow Process

The continuous flow model process simulates conventional activated sludge treatment, which is a widely used biological wastewater treatment process. A series of experiments was performed to explore ways to enhance the population level of GEMs surviving in activated sludge.

First, the basic survival patterns of *E. coli* C600 (pBH500) and *P. putida* BH (pBH500) in the continuous flow process were investigated. The survival tests were carried out at an SRT of 15 days. Cells of the GEMs were harvested by centrifugation and added directly to the aeration tank after being mixed with an appropriate amount of activated sludge mixed liquor. From the time the cells were added, the wastewater feed was interrupted for 24 hours to completely mix the inoculated GEMs with the activated sludge. The result for *E. coli* C600 (pBH500) is shown in Figure 7.4, and that for *P. putida* BH (pBH500) is shown in Figure 7.5.

E. coli C600 (pBH500) inoculated at 1.6×10^6 cfu/mL in the aeration tank of the continuous flow process showed an abrupt decline to 7.8×10^4 cfu/mL in 2 days, followed by a gradual decline of the population. During this period, a stable population of approximately 10^3 to 10^4 cfu/mL was maintained between day 11 and day 35. Thus, a survival time course consisting of a declining phase and a stable phase was also observed in the continuous flow process. The GEM, *E. coli* C600 (pBH500), was detected also in the effluent from the sedimentation tank at the level of 10^1 to 10^2 cfu/mL.

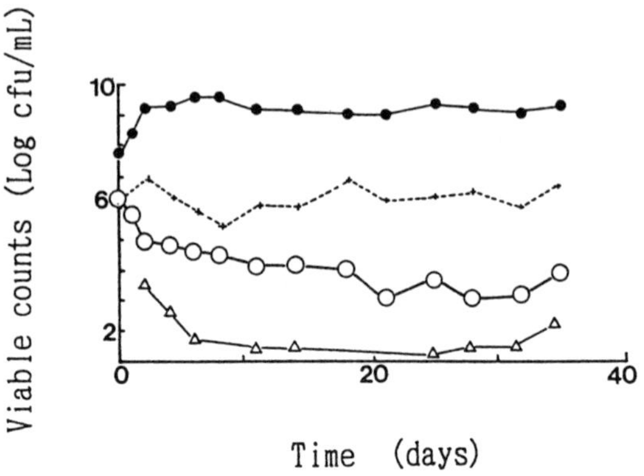

FIGURE 7.4. Survival of *E. coli* C600 (pBH500) in the continuous flow activated sludge process: ●, total bacterial count; ○, C600 (pBH500) in the aeration tank; +, total bacterial count; △, C600 (pBH500) in the effluent.

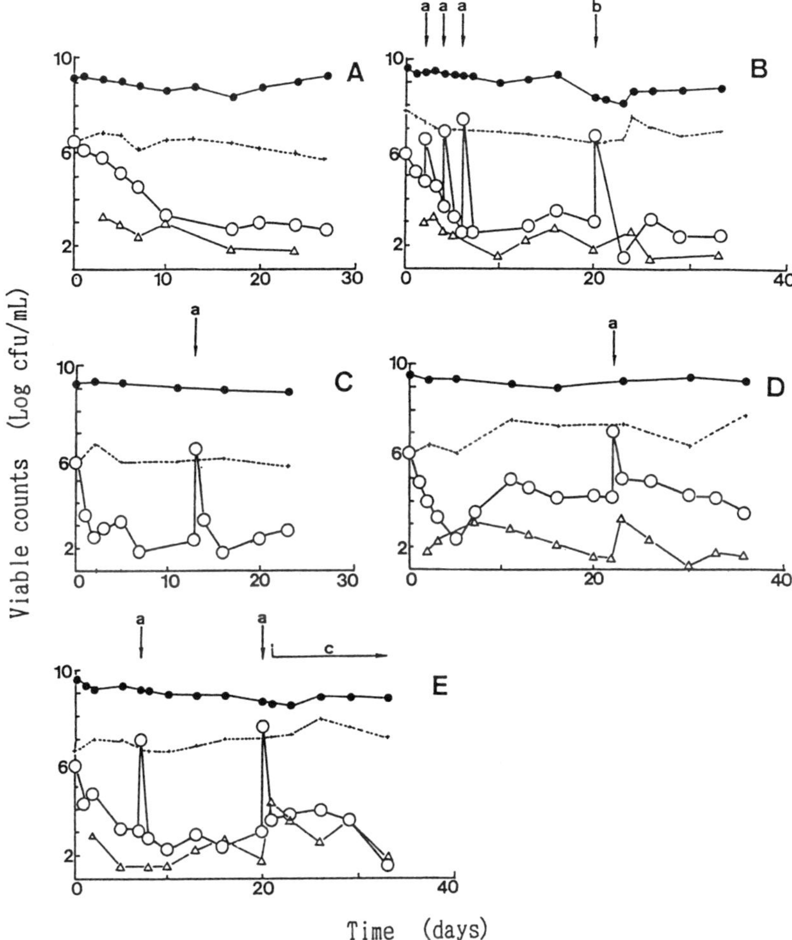

FIGURE 7.5. Survival of *P. putida* BH (pBH500) in the continuous flow activated sludge process: (A) run 1, (B) run 2, (C) run 3, (D) run 4, (E) run 5; ●, total bacterial count; ○, BH (pBH500) in the aeration tank; +, total bacterial count; △, BH (pBH500) in the effluent. a, repeated inoculation followed by 1-hour feed interruption; b, repeated inoculation followed by 24-hour feed interruption; c, addition of phenol to the wastewater.

The basic pattern of the survival of *P. putida* BH (pBH500) in the aeration tank is shown in Figure 7.5(a). In this experiment, namely Run 1, *P. putida* BH (pBH500) was inoculated at 4.4×10^6 cfu/mL. The population of *P. putida* BH (pBH500) showed a rapid drop to 1.8×10^3 cfu/mL in 10 days, followed by a stable phase (at approximately 10^3 cfu/mL as well as the case of *E. coli* C600 (pBH500).

From these two results, it was clarified that the survival of GEMs in the continuous flow process also consists of the initial declining phase and the following stable phase, similar to that observed in the fill-and-draw process, suggesting this survival pattern may be common in activated sludge microcosms.

An additional four (five in all) experiments were carried out to discover the operational parameters affecting the survival of the GEMs in the continuous flow process by using *P. putida* BH (pBH500). Results of the other survival experiments (Run 2 through Run 5) are compared in Figures 7.5(b)–7.5(e) with Run 1.

In Run 2, the inoculation of the GEMs was repeated on day 2, day 4, day 6, and day 20 in addition to day 0 [Figure 7.5(b)]. It is apparent that the spiked GEM population dropped extremely fast to the original levels in a day. Although the inoculum size affected the survival of the GEMs in the fill-and-draw process, the repeated inoculation, which may substitute for a large amount of inoculation, did not enhance GEM survival in the continuous process. In this run, the interruptions of the wastewater feed following the repeated inoculations on day 2, day 4, and day 6 were only 1 hour. The rapid GEM decline immediately after the repeated inoculation was at first considered due to insufficient mixing of the GEMs with the activated sludge flocs. However, although the fourth inoculation was followed by a 24-hour interruption of feed, there was no observed enhancement of GEM survival.

Besides Run 1, in which the SRT was controlled at 15 days, the SRTs were set at 30 and 5 days in Run 3 and Run 4, respectively [Figures 7.5(c) and 7.5(d)], and its effect on the survival of *P. putida* BH (pBH500) in the continuous flow process was investigated. Although the rapid decline of the GEM population in the initial period was observed regardless of the SRT, the population size in the stable phase was affected considerably by SRT control. The stabilized population sizes at SRTs of 5, 15, and 30 days were approximately 10^4, 10^3, and 10^2 cfu/mL, respectively. Thus, an increase in the ecological stability of the GEM with a decrease in the SRT was observed in the continuous flow process as well as in the fill-and-draw process, suggesting a significant effect of the SRT on GEM survival in activated sludge. However, repeated inoculation did not raise GEM survival in Run 3 or Run 4.

Substrates, which can be utilized specifically by introduced GEMs or particular indigenous bacterial populations, seem to affect the survival of GEMs in activated sludge microcosms. Therefore, phenol, which supports the growth of *P. putida* BH (pBH500) as a sole carbon and energy source, was fed from day 21 in the synthetic wastewater at a concentration of 100 mg/L (the GEM was reinoculated on day 20), and its effects on the survival of the GEM was investigated in Run 5 [Figure 7.5(e)]. As shown in the figure, no distinct change in the population of *P. putida* BH (pBH500) occurred.

Through this series of experiments, *P. putida* BH (pBH500) was detected in the effluent from the sedimentation tank at approximately 1/100 the density of that in the aeration tank, with the exception of Run 3 in which the GEM in the effluent was considered to be less than 10^0 cfu/mL, the detectable limit of the plate count method used in this study.

7.2.4 Effects of Operational Parameters

The survival studies of GEMs in both fill-and-draw and continuous flow activated sludge processes showed that GEMs introduced into the sludge microcosm did not disappear, but could maintain their population at relatively stable levels after the rapid decline immediately after inoculation. This suggests the possibility of utilizing the useful abilities of GEMs in activated sludge processes. However, the stable population size of GEMs was considered low. For example, a maximum of the stable population sizes of *E. coli* C600 (pBH500) and *P. putida* BH (pBH500) in the fill-and-draw process was approximately 10^6 to 10^7 cfu/mL, and that in the continuous flow process was approximately 10^4 cfu/mL, which accounted for less than 0.1% of the total bacterial counts of the activated sludge. It seems desirable to establish strategies to maintain them at higher levels. Accordingly, it is also important to elucidate the mechanisms and/or determine the influential parameters concerning the survival of GEMs in activated sludge processes.

In general, the increase or decrease of an introduced GEM in the aeration tank depends on its input and output into or from the tank and its growth and decay.

$$\text{Rate of increase (accumulation) of an introduced GEM} = \text{input} - \text{output} + \text{growth} - \text{decay} \quad (7.1)$$

In the above-mentioned survival experiments, the input of GEMs into the activated sludge process was zero except for the initial inoculation (and repeated inoculation if present), and the output from the process was mainly due to the definite volume of sludge wastage controlled according to the SRT (except when GEMs were contained at a considerable density in the effluent). In the initial day or two of the fill-and-draw process, the turbidity of settled supernatant (effluent) probably was due to free GEMs that were not taken into the activated sludge flocs. Therefore, Equation (7.1) can be expressed in these cases:

$$\text{Rate of increase} = \text{growth} - \text{decay} - \text{output due to sludge wastage} \quad (7.2)$$

$$V(dX/dt) = V\mu X - Vk_d X - Q_w X \quad (7.3)$$

where X is the density (mg/L or cfu/mL, etc.) of GEMs in the aeration tank, t is time, V is volume of the aeration tank, Q_w is the volumetric flow rate of the sludge wastage, μ is the specific growth rate of GEMs, and k_d is the specific decay rate of GEMs in the process.

By multiplying Equation (7.3) by the inverse of the GEM population (mass) present in the aeration tank, $1/VX$, Equation (7.4) is obtained.

$$(dX/dt)/X = \mu - k_d - Q_w/V \quad (7.4)$$

From Equation (2.5) and the definition of SRT, Equation (7.4) becomes:

$$(dX/dt)/X = \mu - k_d - 1/\text{SRT} \quad (7.5)$$

The GEM survival slopes which appeared in the activated sludge processes (Figures 7.1 to 7.5) theoretically express the specific rate of GEM increase, $(dX/dt)/X$. The theoretically expected specific decrease rate of GEMs due to sludge wastage, 1/SRT, is also diagramed in the figures. The comparison of the survival curves, therefore, may indicate differences in the growth and decay rate of GEMs in the activated sludge processes under different conditions.

As shown in the figures, the rapid decline in the GEM population observed in the declining phase cannot be accounted for only by sludge wastage (1/SRT) in all cases, even if the GEMs could not grow at all ($\mu = 0$). This suggests the presence of other strong selective pressures (k_d) which accelerate the decay of GEMs in the activated sludge microcosms, which may include predation by protozoa, competition with indigenous bacterial populations, and fatal environmental effects, i.e., selective pressure of the ecosystem. In the declining phase, the selective pressure of an ecosystem disrupted by the introduction of GEMs is considered high. On the other hand, in the stable phase, it is considered that introduced GEMs have become settled as members of a relatively stable ecosystem of activated sludge, and that the selective pressure was low. The growth (μ) of GEMs in the process seemed to be nearly equal to the ecological selective pressure (k_d) plus wastage (1/SRT) in the stable phase.

According to this study, parameters affecting the survival of GEMs in activated sludge processes are the operation mode (whether the fill-and-draw or the continuous flow mode) and SRT.

GEMs survived at higher levels in the fill-and-draw process than in the continuous flow process. Moreover, the inoculum size considerably affected the survival of GEMs in the fill-and-draw process, while the repeated inoculation had no effects in the continuous flow process. One reason for the differences in GEM survival under the two distinct operation modes may be differences in nutritional conditions. For example, TOC concentration in the fill-and-draw process (flask) showed a change from approximately 200–20 mg/L daily, whereas that in the continuous flow process (aeration tank) was maintained at 10–20 mg/L constantly. Such differences may cause different growth of GEMs (the mean value of μ). Another putative reason seems to be the difference in the physical characteristics of activated sludge flocs, which may affect the taking up of GEMs into the floc microbial community. The flocs formed in the fill-and-draw process were much compacter than those observed in the continuous flow process as a whole.

The SRT also influenced the survival of the GEMs regardless of the operation mode. As the SRT decreased, the survival level of the GEMs increased. The SRT control at smaller values may confer certain advantages in the growth and/or decay on the GEMs. Since the diversity of bacterial flora in the activated sludge increased with the increase in the SRT of the process (Section 2.3), it is considered that the diversity of indigenous bacterial flora might influence the survival (μ and k_d) of GEMs.

On the other hand, phenol did not act as selective pressure with *P. putida* BH (pBH500). This suggests that phenol as a specific substrate is not sufficient to accelerate the growth of *P. putida* BH (pBH500) in a complex ecosystem of mixed substrates and microorganisms such as that in the activated sludge process.

Not only operation mode and operational parameters, but also the physiological properties of GEMs applied to the activated sludge process may be an important factor. McClure et al. (1991) emphasized the importance of choosing bacterial strains which are well-adapted to the actual treatment conditions for the successful use of GEMs or actual bacteria in the wastewater treatment. In their research, a 3-chlorobenzoate(3CB)-degrading GEM, *P. putida* UWC1 (pD10), was introduced into a model activated sludge unit treating 3CB-containing wastewater, and its survival was monitored. Then the fate of two naturally occurring 3CB-utilizing bacterial strains derived from the activated sludge microflora of the unit were reinoculated and monitored. Both strains AS2 and *P. putida* ASR2.8, which seemed better adapted to the environmental conditions of the process, survived longer than *P. putida* UWC1 (pD10).

In our study, two host strains, *E. coli* C600 and *P. putida* BH, were used for GEM breeding. *Pseudomonas* is considered a dominant genera in activated sludge, whereas enteric bacteria, including *E. coli,* have rarely been detected. Interestingly, although *P. putida* BH (pBH500) seemed to be more adaptable to the activated sludge process than *E. coli* C600 (pBH500), the survival or ecological stability of the two strains did not greatly differ.

7.2.5 Conclusions and Considerations

GEMs introduced into activated sludge maintained certain population sizes after abrupt decline, suggesting the possibility of successful uses of GEMs in wastewater treatment process even when the GEMs are directly inoculated. However, before we can realize such applications, it is desirable to elucidate the mechanisms of the survival of GEMs in the activated sludge process so as to establish strategies for enhancing their survival. It is especially important to find the reason(s) for the initial rapid drop in GEM population. The components concerning the selective pressure of the ecosystem in which GEMs are introduced as to be made clear. Basic research on microbial ecology, such as the studies described in Chapter 2, are, therefore, very significant.

From the practical viewpoint, two operational parameters affecting GEM survival were determined—the operation mode and the SRT. A fed batch or sequential batch activated sludge process under short SRT control is considered suitable for the microbial inoculation of GEMs.

Moreover, further studies on the expression and transfer of recombinant genes, an assessment of hazardous effects on the indigenous microflora, etc. may be necessary.

7.3 IMPROVEMENT OF ECOLOGICAL STABILITY

7.3.1 Inoculation of Multiple GEMs

GEMs capable of degrading xenobiotic compounds must be maintained at high population levels for efficient wastewater treatment, because the degradation activity seems to depend upon the population size of the corresponding degrading bacteria. For example, McClure et al. (1991) estimated a minimum population size of the 3-chlorobenzoate degrading GEM commensurate with the catabolism of the substrate within a model activated sludge unit at 10^6 cfu/mL. However, although the GEMs, *P. putida* BH (pBH500) and *E. coli* C600 (pBH500), introduced into the activated sludge microcosms could maintain their populations stably for more than one month after the abrupt decline, the maximum stabilized population size of the GEMs in the survival test using the continuous flow process was approximately 10^4 cfu/mL, and seemed too small to utilize their degradation activity.

Therefore, we propose inoculating multiple GEMs carrying the same recombinant gene(s) as a more effective inoculation strategy for enhancing the survival of the ecological stability of the desired recombinant gene(s) in wastewater treatment.

By using various vectors (see Section 4.1), a desired degrading gene can be introduced into a wide variety of bacterial hosts. When the resultant GEMs carrying the same degrading gene are inoculated together into the microbial community of a wastewater treatment process, and more than two of them are able to maintain their populations at certain stable levels, the number of the surviving GEMs harboring the desired recombinant gene may be higher than when each of the GEMs is inoculated individually.

For example, assuming that two types of GEMs, GEM1, and GEM2, carrying the same recombinant gene (*gem*) are inoculated into the wastewater treatment process and that the GEM1 and GEM2 can maintain their population stably at $X1$ and $X2$ (cfu/mL, etc.) respectively when inoculated separately (separate inoculation), and at $Xm1$ and $Xm2$ respectively when inoculated together (mixed inoculation), the *gem*-carrying population surviving in the process—i.e., the ecological stability of the gene—can be expressed as follows:

- in the separate inoculation: $X1$ or $X2$
- in the mixed inoculation: $Xm1 + Xm2$

Thus, if $(Xm1 + Xm2)$ is greater than $X1$ or $X2$, then the survival level of the recombinant gene has been enhanced by the plus effect of the mixed inoculation. Whether $Xm1(Xm2)$ becomes greater, smaller than, or equal to $X1(X2)$ may depend on the form of the interaction between GEM1 and GEM2, on the assumption that the mixed inoculation has little or no effect on the microbial community of the wastewater treatment process. If both strains lack the interaction, $Xm1(Xm2)$ may be equal to $X1(X2)$, whereas the mutualism (each member benefits from the other) will result in $Xm1(Xm2)$ is being greater than $X1(X2)$. The competition, commensalism, or amensalism between the GEM1 and GEM2 may or may not cause a reverse effect.

Whether such strategy for improving the survival of the recombinant gene can be implemented or not was assessed with a survival test in activated sludge using the fill-and-draw process. In addition to *E. coli* C600 (pBH500) and *P. putida* BH (pBH500), *P. putida* KT2440 (pBH500) was constructed by transforming the recombinant plasmid pBH500 containing the *pheB* gene into another *P. putida* host strain, KT2440, and used in the survival study.

E. coli C600 (pBH500) and *P. putida* BH (pBH500) were inoculated together into the fill-and-draw activated sludge process (the SRT was controlled at approximately 20 days) and the survival of both GEMs was monitored [Figure 7.6(a)]. The survival course of each strain inoculated individ-

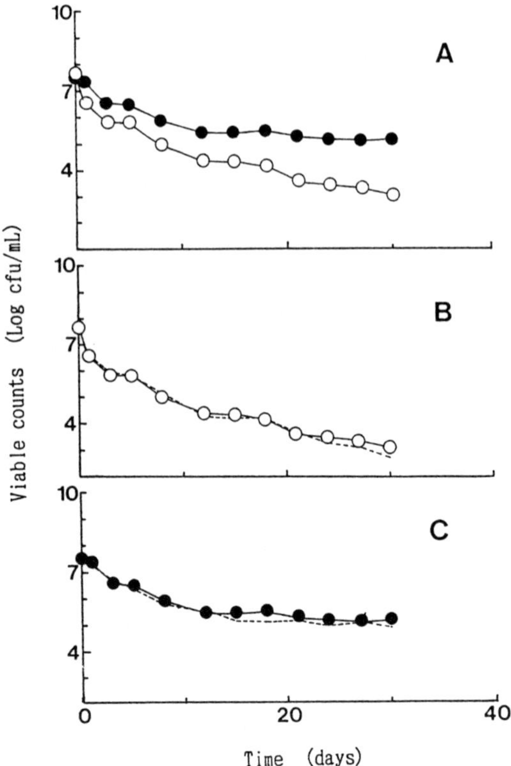

FIGURE 7.6. Survival of *E. coli* C600 (pBH500) and *P. putida* BH (pBH500) in the mixed inoculation: (A) survival of C600 (pBH500) and BH (pBH500) in the mixed inoculation; (B) survival of C600 (pBH500) in the separate inoculation; (C) survival of BH (pBH500) in the separate inoculation ○, C600 (pBH500); ●, BH (pBH500). Dotted lines in (B) and (C) are survival courses in the mixed inoculation.

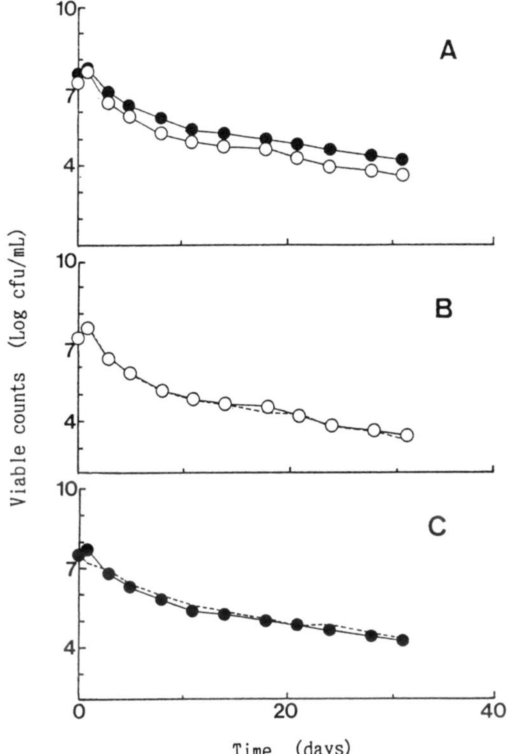

FIGURE 7.7. Survival of *P. putida* BH (pBH500) and *P. putida* KT2440 (pBH500) in the mixed inoculation: (A) survival of BH (pBH500) and KT2440 (pBH500) in the mixed inoculation; (B) survival of BH (pBH500) in the separate inoculation; (C) survival of KT2440 (pBH500) in the separate inoculation ○, BH (pBH500); ●, KT2440 (pBH500). Dotted lines in (B) and (C) are survival courses in the mixed inoculation.

ually was also shown and compared in the figure [Figures 7.6(b) and 7.6(c)]. As shown in the figures, the survival courses of *E. coli* C600 (pBH500) and *P. putida* BH (pBH500) in the mixed inoculation and the separate inoculation showed little or no differences, suggesting little or no interactions between the strains.

Another survival study using genetically engineered *P. putida* BH (pBH500) and *P. putida* KT2440 (pBH500) exhibited similar results, as shown in Figure 7.7. It was assumed that both GEMs had very similar physiological properties because they belonged to the same species *P. putida*, so certain biological interactions, such as competition, should be observed between these two GEMs when inoculated together into the activated sludge. However, the results suggested no interactions.

From these experimental results, the plus effect of the mixed inoculation of GEMs harboring the *pheB* gene was recognized. For example, in the experiment using *P. putida* BH (pBH500) and *P. putida* KT2440 (pBH500), the stable *pheB*-carrying population surviving in the activated sludge on day 20 was 2.4×10^4 cfu/mL [*P. putida* BH (pBH500)] or 6.2×10^4 cfu/mL [*P. putida* KT2440 (pBH500)] when they were inoculated individually, while it was the sum of 2.4×10^4 cfu/mL of *P. putida* BH (pBH500) and 6.3×10^4 cfu/mL of *P. putida* KT2440 (pBH500), equal to 8.7×10^4 cfu/mL, when inoculated together. Thus, it may be concluded that the survival of the *pheB*-carrying strains was improved by this strategy.

7.3.2 Immobilization of GEMs

Recently, much research into the application of immobilization techniques to wastewater treatment have been carried out, and their high potential for improving the treatment performance and solving problems concerning solid-liquid separation in the settling tank has been recognized.

A most important advantage of the use of immobilization techniques in wastewater treatment is to be able to maintain high biomass (activated sludge, etc.) concentration, which is regarded as impossible to attain in a normal "suspended" biological process. This advantage also seems useful in maintaining high concentrations of stable GEMs in the wastewater treatment process. Moreover, since the GEMs can be physically separated from the microbial community of the wastewater treatment process when immobilized, the selective pressure of the ecosystem, including predation by protozoa and competition with indigenous bacterial populations (see Section 7.2), may be reduced. Therefore, the ecological stability of the GEMs should improve. The GEM, *P. putida* BH-M21 (pHF400), was immobilized by the PVA (polyvinyl alcohol)-boric acid method (Hashimoto and Furukawa, 1987), and the immobilized GEM beads were added to the activated sludge for evaluating the effects.

P. putida BH-M21 is a derivative of the phenol-degrading bacterium, *P. putida* BH, which requires cystine and glycine for growth. The GEM was constructed by introducing the recombinant plasmid, pHF400, carrying *nahG* into *P. putida* BH-M21. The resulting GEM can degrade phenol and salicylate simultaneously, as well as *P. putida* PpG1064 (pHF400) which was described in Section 5.2.

The PVA-boric acid method used for immobilizing the GEM is characterized by gelling of the PVA-bacterial cell mixture in a saturated boric acid solution, and by the entrapment of the bacterial cells in a monodiol-type PVA-boric acid gel lattice. This polymerization reaction is expressed as follows:

$$(-CH_2-\underset{\underset{OH}{|}}{CH}-CH_2-\underset{\underset{OH}{|}}{CH}-)_n + nH_3BO_3 \rightarrow (-CH_2-\underset{\underset{O}{|}}{CH}-CH_2-\underset{\underset{O}{|}}{CH}-)_n \text{ (O–B(OH)–O bridge)} \tag{7.6}$$

This reaction is an instantaneous polymerization, so spherical immobilized cell beads are easily formed. Compared to commonly used polymeric substances such as acrylamide, $\varkappa$-carrageenan, calcium alginate, and agar, this PVA-boric acid method seems suitable for wastewater treatment because the cost of chemicals required for immobilization is lower, and the gel is stronger.

The GEM grown in LB broth was centrifuged, and the cell pellet containing dry cell at approximately 100 g/L was prepared for immobilization. One portion of this cell pellet was mixed thoroughly with one portion of 20% PVA (PVA-HC–100% saponification, 2,000 degrees of polymerization) aqueous solution. This mixture was dropped into a gently stirred saturated boric acid solution to form spherical beads, which were kept in the solution for approximately 24 hours under gentle stirring to complete gelation inside the beads. Then the beads were washed with water, reactivated in the minimal salts medium supplemented with phenol, salicylate (each at 300 mg/L), cystine, and glycine for more than one day, and then used in the experiment.

Either suspended cells or entrapped-immobilized cells of the GEM (suspended GEM and immobilized GEM, respectively) were introduced into the activated sludge. The phenol and salicylate degradation by such activated sludge-GEM mixture was investigated in batch cultures. Activated sludge grown with the synthetic wastewater (containing mainly peptone and meat extract) and the suspended or immobilized GEMs precultured with the medium containing phenol and salicylate were inoculated at 13 and 10 mg/L, respectively, to 250 mL of the synthetic wastewater containing phenol and salicylate in a 500-mL Erlenmeyer flask, then the flasks were rotated on a rotary shaker at approximately 120 rpm at 30°C.

Results were shown in Figure 7.8. In the figure, the degradation properties of phenol and salicylate by the activated sludge containing *P. putida* BH-M21 (pHF400) as the suspended cells and as the immobilized cells, were compared with that of the activated sludge without the GEM. Phenol degradation activity was enhanced slightly by adding the GEMs, while the degree of the enhancement depended only slightly upon whether the cells of the GEM were immobilized or not. On the other hand, salicylate was removed a little faster by the activated sludge with the suspended GEM, and

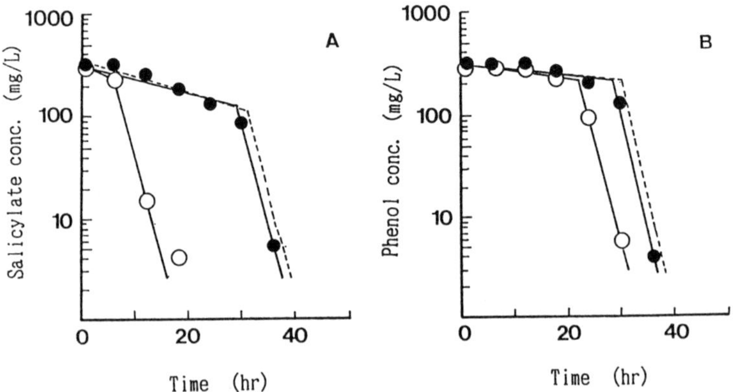

FIGURE 7.8. Salicylate and phenol degradation by immobilized GEM in the activated sludge: (A) phenol degradation; (B) salicylate degradation. -----, activated sludge without GEM; ●, activated sludge with suspended GEM; ○, activated sludge with immobilized GEM.

much faster by that with the immobilized GEM than by that without the GEM. These results suggest that the GEM *P. putida* BH-M21 (pHF400) was protected from the ecological selective pressure of the activated sludge by immobilization, and consequently that the salicylate degradation activity of the GEM was maintained in the activated sludge. On the other hand, when the GEM was not immobilized, the degradation activity could not be shown fully, maybe due to the presence of microbial interactions with indigenous activated sludge microorganisms. The reason for the slight enhancement in phenol degradation—which was observed even when immobilized GEM was added to the activated sludge—is considered to be a masking effect by relatively high activity of indigenous phenol-degrading bacterial populations.

It has been demonstrated that immobilization techniques can enhance the ecological stability of GEMs in wastewater treatment. However, further long-term practical experiments may be necessary.

7.3.3 Breeding of Floc-Forming GEMs

Although immobilization seems to be one of the easiest and most effective methods for maintaining GEMs in the wastewater treatment process, it is so expensive—even when inexpensive materials such as PVA and boric acid are used—that it would be hard to apply on a large scale. Therefore, we propose that the microorganism which can grow as flocs or which can adhere to the biofilm should be used as recipients to inexpensively improve the ecological stability of GEMs. Floc formation and adhesive growth are considered advantageous properties, enabling microorganisms to remain in suspended-culture processes (activated sludge) and in biological film flow processes (trickling filters). In this section, the breeding of a floc-forming GEM by *in vivo* DNA recombination is described as such an example.

The floc-forming bacterial strain *Pseudomonas lemoignei* 551 used in this experiment was selected from among 199 bacterial strains isolated from activated sludge or biofilm samples (145 strains from four activated sludge samples and 54 strains from two biofilm samples).

First, typical colonies obtained by the plating of samples were chosen, and their ability to form flocs in synthetic wastewater (TOC = 170 mg/L) was investigated by cultivation on a reciprocal shaker at 70 oscillations per minute at 30°C. Floc-forming ability was evaluated by the index of flocculation (IF) defined as follows (Tanaka et al., 1985):

$$\mathrm{IF}(\%) = (OD_t - OD_{sup})/OD_t \times 100 \qquad (7.7)$$

where the OD_{sup} indicates a biomass (cell) concentration of the supernatant of culture broth (in the synthetic wastewater) after 30 min settling in 10-mL graduated cylinder, and the OD_t indicates total biomass concentration of the culture broth. The biomass concentration was measured as OD_{600}. A large IF value means relatively better floc formation and shows a good settling characteristic of the culture.

Thus, six bacterial strains which showed more than 70% of the IF value at 1-day and 3-day cultivation were selected as floc-forming bacteria among the 199 strains. From among these twelve strains, *P. lemoignei* 551, isolated from an activated sludge sample, was selected for this study because:

- *P. lemoignei* 551 is a gram-negative strain, which may be suitable as a host for NAH plasmid that can replicate in a wide variety of gram-negative bacterial strains.
- It could form flocs on various sole carbon sources such as maltose, cellobiose, pyruvate, acetate, succinate, proline, DL-β-hydroxybutyrate, etc. as well as in synthetic wastewater and CGY medium (1/10 strength), while the other strains could grow or show floc formation in fewer media.
- Although floc formation of the other strains deteriorated as the cultivation was prolonged, *P. lemoignei* 551 maintained good floc formation, or setting characteristics, for a relatively long time. For example, the IF value of another floc-forming strain, *Pseudomonas vesicularis* K11, in synthetic wastewater dropped to approximately

60% on day 3 from the maximum value of 83% on day 1, while *P. lemoignei* 551 maintained its IF value at more than 80% from day 1 to day 4.

The genetic engineering methodology used for constructing the floc-forming GEM was the transfer of natural plasmid to the recipient. Conjugation by means of the filter mating method (Bagdasarian et al., 1981) was performed between *P. putida* PpG1064 (NAH) as the donor strain, and floc-forming *P. lemoignei* 551 as the recipient strain. *P. putida* PpG1064 (NAH) and *P. lemoignei* 551 were grown to a cell density that gave an OD_{600} of approximately 1.0 in LB broth, and both cultures were mixed at a ratio of 1:10. This ratio was used because the specific growth rate of the donor strain was much faster than that of the recipient strain. Then, 0.1 mL of the mixture was spread onto a sterile membrane filter (diameter, 10 mm; pore size, 0.45 μm) on LB agar plate. After a 6-hour incubation at 30°C, the filter was transferred into a test tube containing 1 mL of 0.9% NaCl solution, which was vortexed to resuspend the bacterial cells on the filter. The cell suspension was diluted and plated onto a selective medium (containing salicylate as a sole energy and carbon source) for the floc-forming GEMs. As the naturally-occurring plasmid NAH is a self-transmissible plasmid coding for naphthalene and salicylate degradation (Dunn and Gonsalus, 1973), this filter mating was designed to transfer the NAH plasmid from *P. putida* PpG1064 (NAH) to the floc-forming recipient, *P. lemoignei* 551, that was anticipated to result in the construction of the naphthalene or salicylate-degrading and floc-forming transconjugants.

In this filter mating method, transconjugants appeared on the selective medium at a transfer frequency of 5×10^{-6} cfu per donor or 1×10^{-7} cfu per recipient. One of the transconjugants selected showed the ability of floc-formation derived from the recipient strain, and also had the ability to mineralize salicylate and naphthalene derived from the donor strain or the NAH plasmid, as anticipated. Moreover, a plasmid a little smaller than the NAH was detected in it, although no plasmid had been detected in the parental recipient, *P. lemoignei* 551. Therefore, we designated the transconjugant *P. lemoignei* 551 (NAH′).

The floc-forming ability and the salicylate removal of this GEM in CGY broth containing salicylate at 400 mg/L were investigated and compared with those of the parental strains in Table 7.1. Though the salicylate removal of *P. lemoignei* 551 (NAH′) was lower than that of the donor *P. putida* PpG1064 (NAH), it showed flocculent growth as good as the recipient *P. lemoignei* 551.

Since the floc-forming bacteria are expected not to be washed out from the activated sludge process or other suspended-culture processes even if

TABLE 7.1. Breeding of a floc-forming GEM.

Strain	Salicylate Removal (%)	IF (%)
Recipient: *P. lemoignei* 551	0	97.5
Donor: *P. putida* PpG1064 (NAH)	99.0	0
Conjugant: *P. lemoignei* 551 (NAH′)	33.4	97.7

Reprinted from *Water Research, Vol. 25,* Fujita, Ike and Hashimoto, "Feasibility of Wastewater Treatment Using Genetically Engineered Microorganisms," p. 984, 1991, with kind permission from Pergamon Press Ltd., Headington Hill Hall, Oxford OX3 0BW, UK.

they cannot grow fast, it may be concluded that we can strengthen the ecological stability, and degrade xenobiotic compounds continuously, by using a floc-forming and degrading GEM such as *P. lemoignei* 551 (NAH′).

7.4 REFERENCES

Amy, P. S. and H. D. Hiatt. 1989. "Survival and Detection of Bacteria in an Aquatic Environment," *Appl. Environ. Microbiol.*, 55:788–793.

Awong, J., G. Bitton and G. R. Chaudhry. 1990. "Microcosm for Assessing Survival of Genetically Engineered Microorganisms in Aquatic Environments," *Appl. Environ. Microbiol.*, 56:977–983.

Bagdasarian, M., R. Luze, B. Lückert, F. C. H. Franklin, M. M. Bagdasarian, J. Frey and K. N. Timmis. 1981. "Specific-Purpose Plasmid Cloning Vectors, II. Broad Host Range, High Copy Number, RSF1010-Derived Vectors and Host-Vector System for Gene Cloning in *Pseudomonas*," *Gene*, 16:237–247.

Beringer, J. E. and M. J. Bale. 1988. "Survival and Persistance of Genetically Engineered Microorganisms," in *The Release of Genetically Engineered Microorganisms*, M. Sussman, C. H. Colling, F. A. Skinner and D. E. Stewart-Tull, eds., London: Academic Press, pp. 29–46.

Dunn, N. W. and I. C. Gunsalus. 1973. "Transmissable Plasmid Coding Early Enzymes of Naphthalene Oxidation in *Pseudomonas putida*," *J. Bacteriol.*, 114:974–979.

Hashimoto, S. and K. Furukawa. 1987. "Immobilization of Activated Sludge by PVA-Boric Acid Method," *Biotechnol. Bioeng.*, 30:52–59.

Macnaughton, S. J., D. A. Rose and A. G. O'Donnell. 1992. "Persistence of a *XylE* Marker Gene in *Pseudomonas putida* Introduced into Soils of Differing Texture," *J. Gen. Microbiol.*, 138:667–673.

McClure, N. C., J. C. Frey and A. J. Weightman. 1991. "Survival and Catabolic Activity of Natural and Genetically Engineered Bacteria in a Laboratory-Scale Activated-Sludge Unit," *Appl. Environ. Microbiol.*, 57:366–373.

McClure, N. C., A. J. Weightman and J. C. Frey. 1989. "Survival of *Pseudomonas putida* UWC1 Containing Cloned Catabolic Genes in a Model Activated-Sludge Unit," *Appl. Environ. Microbiol.*, 55:2627–2634.

Ramos, J. L., E. Duque and M.-I. Ramos-Gonzalez. 1991. "Survival in Soils of an Herbicide-Resistant *Pseudomonas putida* Strain Bearing a Recombinant Plasmid," *Appl. Environ. Microbiol.*, 57:260–266.

Tanaka, H., N. Kurano, S. Ueda, M. Okazaki and Y. Miura. 1985. "Model System of Bulking and Flocculation in Mixed Culture of *Sphaerotilis* sp. and *Pseudomonas* sp. for Dissolved Oxygen Deficiency and High Loading," *Wat. Res.*, 19:563–571.

CHAPTER 8

Proposal of Wastewater Treatment Processes Using Genetically Engineered Microorganisms

An increasing number of techniques for efficiently utilizing microbial activities in various types of bioreactors or bioprocesses have been developed in recent years. Immobilization of microorganisms or enzymes, ultrafiltration techniques for solid/liquid separation, bioprocesses equipped with automatic controllers, etc., are included in such techniques. Although these techniques may be applicable to the use of GEMs in wastewater treatment processes, the possibilities have not been discussed fully.

Taking the application of these techniques into consideration, some possible wastewater treatment processes using GEMs are proposed and evaluated in this chapter based on the knowledge, theories, and experimental research concerning the breeding and application of GEMs described in the previous chapters. Direct inoculation of GEMs into the wastewater treatment microcosms, constant supply of GEMs by feeding systems, and pretreatment of hazardous components of wastewater with pure cultures of GEMs are discussed here.

8.1 INOCULATION OF GEMs

To inoculate GEMs directly into the mixed microbial flora of wastewater treatment processes, such as activated sludge or biofilm, in the start-up periods is the most simple way to utilize their abilities or functions.

If the GEMs inoculated into the microcosm are able to maintain their population at a sufficiently high level to work (such as degradation of xenobiotic compounds, detoxification of hazardous wastes, and so on) and, consequently, to contribute to the improvement of the treatment performance of the processes, this method may be the most ideal, especially from the viewpoint of economy, since no modifications of the existing processes are required.

The successful use of GEMs in this way depends upon both genetic and ecological stability of GEMs in the actual treatment processes. In other words, the breeding or the selection of stable GEMs is essential.

On the other hand, operational modes and parameters of the treatment processes are also important factors, since they may affect the survival of GEMs in the wastewater treatment microcosms. For example, the fill-and-draw and sequential batch systems are considered more suitable for GEM applications than the conventional continuous flow systems, as described in Section 7.2.

GEMs which exhibit extremely high degradation activities of xenobiotic compounds may be suitable for use in this way, since the degradation rate is considered to be enhanced even if the GEMs maintain relatively small populations in the processes. The floc-forming GEM seems to be another suitable GEM for use in activated sludge processes because of its high ecological stability. Moreover, the application of immobilization techniques may also enhance the ecological stability of GEMs in the processes as described in Section 7.3.2.

8.2 GEM-SEEDING SYSTEM

It is impossible or very difficult to consistently maintain selective pressure for specific bacterial strains, including GEMs, in actual wastewater treatment processes, since the feed of wastewater fluctuates both qualitatively and quantitatively, and the environmental conditions of the treatment cannot be controlled easily. Therefore, in cases where GEMs inoculated into the microcosms of the processes are not able to survive at a sufficient level for enhancing the treatment performance, it is necessary to supply the GEMs as additional inoculant in order to maintain a sufficient population for the treatment processes.

An example of the modified activated sludge process for this strategy is diagrammed in Figure 8.1. This process is a conventional activated sludge process equipped with the *GEM-seeding system* which supplies the main treatment reactor (aeration tank) with cells of GEMs prepared by a certain method. If the GEMs decline rapidly in the treatment process, the GEMs should be added continuously, while intermittent inoculation may be possible in cases where the GEMs decline gradually and, consequently, can maintain sufficiently large populations for relatively long periods.

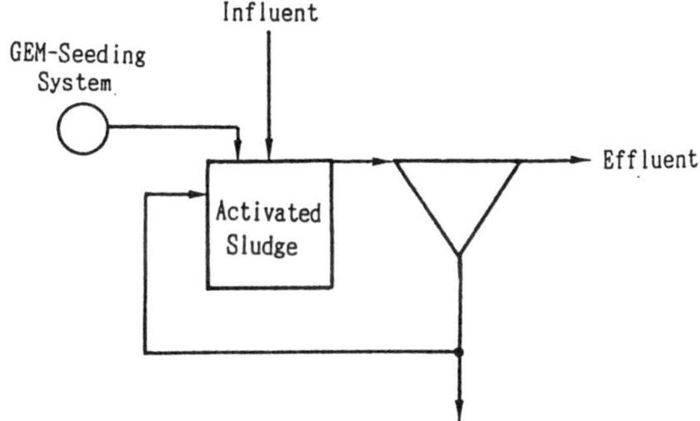

FIGURE 8.1. A GEM-seeding system for activated sludge processes.

A similar microorganism-seeding process, not for GEMs but for naturally occurring degrading bacteria, was proposed and experimentally studied by Cardinal and Stenstrom (1991). Their research was designed to enhance the removal of polynuclear aromatic hydrocarbons, which are a group of significant hazardous compounds contained in combined domestic/industrial wastewater, by a modification of the conventional activated sludge process. In the proposed system, designated *enricher reactor system*, a second biological reactor (an enricher reactor) was used to grow bacterial cells especially prepared to degrade the target xenobiotic compounds, naphthalene and phenanthrene. The enricher reactor was operated as a sequencing batch reactor, and the feed for the enricher contained salicylate and traces of naphthalene as enrichment substrates to induce the catabolic enzymes of the target compounds effectively. The prepared cells were then added to the continuous flow main treatment reactor (aeration tank) as a form of *bioaugmentation,* that is to increase the population of degrading bacteria in the reactor. This new system showed enhanced removal of the target compounds when compared to a control activated sludge process without addition of degrading microorganisms, and they concluded that it is a promising way to improve existing activated sludge process efficiency.

The system is considered applicable to the seeding of GEMs as well as naturally occurring microorganisms. In the research by Cardinal and Stenstrom, salicylate and naphthalene (which themselves are hazardous compounds) were used for propagating or preparing degrading microorganisms in the enricher reactor, since it is necessary to induce the catabolic enzymes coded on naturally regulated genetic informations in natural microorganisms. However, this technique is undesirable for economic and safety reasons. If we can create and use GEMs capable of degrading target compounds without expensive or hazardous inducers by genetically manipulating the regulatory systems for the catabolic genes, application of such a seeding system may generate more economical and desirable processes.

Preparing the GEMs in a second biological reactor, such as an enricher reactor, separated from the main treatment reactor, then adding them to the main reactor has merit. Separation of the GEM seeding system allows it to be operated under various conditions which may be different from that of the main reactor. Design and operational parameters, such as the mean cell residence time, hydraulic retention time, temperature, composition of the growth medium, rate of influent feed, and oxygen supply can be controlled easily to maintain the genetic stability and/or the expression of cloned genes of the desired GEMs at sufficiently high levels. Sequencing batch cultivation, adopted by Cardinal and Stenstrom, is considered suitable for this application because of its versatility, low maintenance requirements, and ability to handle shock loading.

By using such systems, a wide variety of GEMs which have various physiological properties may be utilized in wastewater treatment processes. However, before the concept of GEM seeding systems can be put to use, the problems of cost for the growth substrates and the prevention of microbial contamination of the GEM culture must be solved.

8.3 USE OF PURE CULTURE OF GEMs

To avoid the problem of ecological instability in the mixed microbial flora of wastewater treatment processes, we propose the use of GEMs as a pure culture.

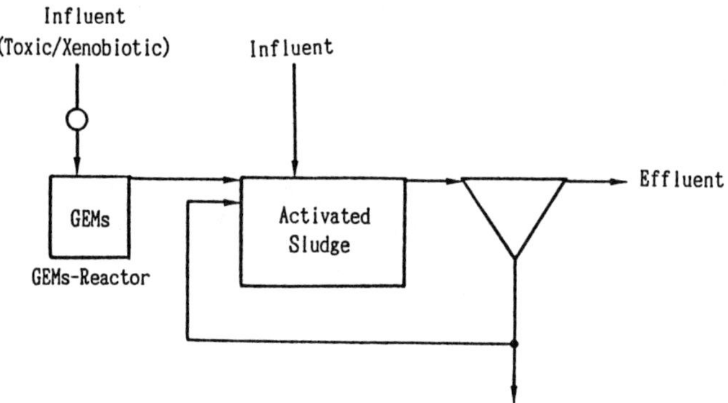

FIGURE 8.2. Use of pure culture of GEMs in activated sludge processes.

An example of such an application of GEMs in the activated sludge process is diagrammed in Figure 8.2. This system is designed to treat influent composed of wastewater from both domestic and industrial origins, and it is assumed that the wastewater from industrial origin contains hazardous compounds which may cause the deterioration of treatment performance in the existing processes. Wastewater from domestic and industrial origins are collected separately, and the former is conducted to the main reactor (aeration tank), while the latter is first conducted into the GEM reactor (pure culture). The hazardous compounds are eliminated in the GEM reactor and the effluent from it is subsequently conducted into the main reactor for complete treatment.

This concept is similar to GEM seeding systems in that the GEMs are cultivated in a second reactor separated from the main treatment reactor or the indigenous microbial communities, whereas it is different in that the wastewater is directly fed to the GEMs, and no other growth substrates are supplied. Therefore, this concept is considered less expensive than GEM seeding. However, since the influent to the GEM reactor (industrial wastewater) may fluctuate qualitatively and quantitatively, stable maintenance of treatment by the pure culture of GEMs requires higher skills and better management techniques for microbial cultivation.

GEM reactors could not only improve the removal efficiency of the target compounds but also reduce the cost of pretreatment for hazardous wastes contained in industrial discharges conducted to the public wastewater treatment facilities, and protect them from upsets caused by the shock loading or uncontrolled discharge of such troublesome compounds.

8.4 REFERENCE

Cardinal, L. J. and M. K. Stenstrom. 1991. "Enhanced Biodegradation of Polyaromatic Hydrocarbons in the Activated Sludge Process," *Res. J. Wat. Pollut. Cont. Fed.*, 63:950–957.

CHAPTER 9

Future Prospects

The technical and social aspects of GEMs in wastewater treatment processes will be discussed in this chapter. A wide variety of technical aspects were described in previous chapters. Those aspects are summarized to show a direction of development for GEM techniques. The most important social issues concerning this approach, the regulatory aspects for the intentional (or accidental) release of GEMs in the environment, are examined and the possibility of real uses of GEMs in wastewater treatment are discussed from such aspects. Finally, the question of whether or not GEM techniques can be applied or not is addressed.

9.1 TECHNICAL PROBLEMS

Technical problems which have to be overcome before GEMs can be used in wastewater treatment are summarized below.

Difficulties in the breeding of GEMs applicable to actual wastewater treatment processes (Chapter 4, Chapter 5). Although desired genes are introduced into host strains by genetic manipulation, a "perfect" or "complete" GEM cannot be constructed in most cases because of various barriers to the efficient expression of foreign genes in the hosts, and unexpected adverse effects caused by genetic manipulation upon the original functions of the hosts.

Genetic instability of constructed GEMs under real treatment conditions (Chapter 6). Foreign genes are unstable in host strains by nature, especially when they are introduced into hosts with plasmid vectors. Two major genetic instabilities of recombinant plasmids—segregational and structural—are commonly observed in GEMs cultured in the absence of certain selective pressures, which cannot be maintained constantly and effectively in the actual wastewater treatment processes. GEMs which cannot express recombinant genes because of genetic instability, of course, will not contribute to the improvement of the treatment, although they can survive in the processes.

Insufficient survival of inoculated GEMs in the microbial communities of wastewater treatment processes (Chapter 7). Successful uses of GEMs in natural environments, such as wastewater treatment processes, depend on their survival in mixed microbial communities at a level high enough to allow sufficient expression of cloned genes. In general, however, it is believed that natural ecosystems exhibit adverse selection on introduced microbial members. Consequently, GEMs which are inoculated into the wastewater treatment processes may not be able to maintain their population sufficiently and stably.

However, these problems involving technical or scientific aspects may be solved to a considerable extent by various countermeasures described in the corresponding chapters.

Difficulties in the breeding of GEMs may be overcome by utilizing versatile cloning vectors and bacterial hosts. Broad host range vectors described in Section 4.2 are good examples of such versatile materials used in genetic manipulations, allowing the introduction of a desired foreign gene into a wide variety of microbial populations, which may include a host capable of expressing the recombinant gene or having very desirable properties for improving the performance of wastewater treatment. Actually, the introduction of the phenol hydroxylase gene *pheA* into an *E. coli* host with a narrow host range vector did not confer detectable phenol degradation activity on the host, while when the *pheA* gene was introduced into *P. putida* hosts with broad host range vectors it enabled them to degrade and grow on phenol efficiently (see Section 4.3). Such versatile materials and methodologies are being developed more and more.

Most currently available vectors for bacteria are based on multicopy plasmids which have no stabilizing mechanisms for their maintenance. As long as we use such plasmid vectors to breed GEMs, the occurrence of plasmid-free segregants (a main genetic instability phenomenon) seems to be inevitable in their application to the wastewater treatment processes because of the lack of selective pressure for the recombinant plasmids. However, the experimental results shown in Section 6.3 suggest the presence of stable combinations of plasmids and host strains. Selection of such GEMs may, to a certain extent, lead to the solution of genetic stability problems. Moreover, transposon vectors which can insert the cloned genes (DNA fragments) into the bacterial chromosome, where they should be maintained as stably as the chromosomal genes, have recently been developed for stable inheritance of the engineered functions in the absence of selective pressure for a vector, and their versatilities are now being improved (Herrero et al., 1990; Barry, 1988).

The last technical problem may be the ecological stability of GEMs. Although GEMs inoculated into microbial communities of wastewater treatment processes seem to survive for a long time, as described in Section 7.2, their population cannot always be maintained at a level high enough to express their activities fully. Section 7.3 provided a few countermeasures for enhancing the survival of GEMs in activated sludge (application of immobilization technique, use of floc-forming hosts for breeding, etc.) and Chapter 8 showed the possible wastewater treatment processes coping with such problems (seeding system of GEMs, use of pure culture of GEMs, etc.). However, none was a final solution. For solving this problem, or for developing strategies for effective GEM inoculation, further studies on the bacterial ecology of wastewater treatment processes are necessary.

9.2 SOCIAL ISSUES

The last problem for the actual uses of GEMs in wastewater treatment may be social issues, which involve regulations and public concern. Social issues must be taken into account when considering the possible uses of GEMs.

Presently, the release of GEMs is more or less restricted. For example, in the U.S., where genetic engineering has been the most developed, and industries have strong desires to utilize this technique commercially, the biotechnology policy statement, *Coordinated Framework of Regulation of Biotechnology, Announcement of Policy and Notice for Public Comment* of the U.S. Office of Science and Technology Policy was issued in 1986, enumerating the following four principles for the regulation of recombinant DNA techniques (Schechtman, 1992).

- The products of recombinant DNA technology will not differ fundamentally from unmodified organisms or from conventional products.
- Existing laws are adequate to regulate the products of the technology.
- Products, not processes, should be regulated, and regulation should be risk-based.
- Regulation should be directed toward the intended end use of products, and should be concluded on a case-by-case basis.

This statement documented the specific role in regulating controlled introductions of GEMs into the environment for the Environmental Protection Agency (EPA), and the EPA stated that microorganisms produced by recombinant DNA, RNA, and cell fusion should be subject to agency review. The EPA also announced that "microorganisms and their DNA molecules are chemical substances" which should be subject to regulations under the Toxic Substances Control Act (TSCA), and defined microorganisms subject to or not subject to reporting.

MICROORGANISMS SUBJECT TO SPECIAL REPORTING

- intergenic microorganisms manufactured or imported for commercial purposes that are not listed on the TSCA inventory
- intergenic microorganisms manufactured or imported for research and development that are released
- pathogens or microorganisms with pathogenic material

MICROORGANISMS NOT SUBJECT TO REPORTING

- naturally occurring microorganisms
- genetically modified microorganisms other than intergenic microorganisms resulting only from addition of well-characterized, noncoding regulatory regions
- contained microorganisms at the research and development level

Countries of the European Community and Japan are in the process of finalizing their guidelines for the release of GEMs in the environment.

The need for such regulations arose from the potential risks caused by GEM release:

- GEMs harboring recombinant genes may exhibit unknown pathogenicity to humans, other animals, and plants.
- GEMs or secondary hosts for recombinant genes which are transferred by natural processes in the environments could have adverse effects upon ecosystems.

Of the two possible risks, pathogenicity of GEMs, is considered to depend on the strains and genes used in their construction. Therefore, as long as we construct GEMs by using well-known or analyzed safe materials through appropriate methodologies, pathogenicity may be avoidable. For example, it is not likely that a newly constructed GEM would produce a pathogenic material which is not produced by its unmodified parental strain.

On the other hand undesirable ecological effects of introduced GEMs cannot be assessed easily, and concern has focused on that. The introduction of a GEM could alter the balance among indigenous species and could cause the following possible undesirable effects, which were reviewed by Pickup et al. (1991).

- extreme fluctuations in microbial population size
- simplification of ecosystem structure or reduction in species diversity
- elimination of critical species involved in key processes
- reduction in primary and/or secondary productivity of ecosystem
- interference with normal nutrient cycling and energy flow

Thus, from the regulatory viewpoint, GEM release in the environment requires a deep comprehension of the behavior of GEMs or the recombinant DNA sequences and their effects on natural ecosystems, including methodologies necessary for such investigations. At any rate, a GEM should not be released into the environment until it is considered safe by both regulators and scientists/technologists.

Since the application of GEMs to wastewater treatment is one of the forms of GEM release, possible risks should be assessed carefully, of course, before it is done. However, although they cannot be assessed fully, the possibility of such risks caused by GEM application in wastewater treatment is considered to be much less than that in other environmental uses, such as the *in situ* bioremediation of polluted sites, because:

- In general, it is unlikely that foreign genes specifying animal or plant pathogenicity are introduced into microorganisms designed to improve wastewater treatment.
- As GEMs are used under controlled conditions in wastewater treatment, their behavior can be anticipated and managed with relative, but not complete, certainty.
- Physical containment of a GEM in the wastewater treatment plant can be achieved to a certain degree by the solid/liquid separation systems (e.g., sedimentation, filtration) and by the sterilization systems (e.g., chlorination, ozonation) for removing biomass possibly containing GEMs from the effluent, and for inactivating the remaining microorganisms possibly including GEMs.

9.3 CONCLUSION

Three major technical problems for the improvement of wastewater treatment by GEM application—difficulties in the breeding, genetic instability, and ecological instability of GEMs—will be overcome by selecting appropriate materials and technologies from among a wide variety of available options.

On the other hand, since two types of possible risks, pathogenicity and adverse ecological impacts of GEMs, are considered unlikely in carefully designed treatment processes, regulators may permit our proposal for using GEMs in wastewater treatment in due course, judging from the sympathetic viewpoints of the possible benefits and risks.

Therefore, it may be concluded that advanced wastewater treatment processes using GEMs are expected in the near future.

9.4 REFERENCES

Barry, G. F. 1988. "A Broad-Host-Range Shuttle System for Gene Insertion into Chromosomes of Gram-Negative Bacteria," *Gene,* 71:75–84.

Herrero, M., V. de Lorenzo and K. Timmis. 1991. "Transposon Vectors Containing Non-Antibiotic Resistance Selection Markers for Cloning and Stable Chromosomal Insertion of Foreign Genes in Gram-Negative Bacteria," *J. Bacteriol.*, 172:6557–6567.

Pickup, R. W., J. A. W. Morgan, C. Winstanley and J. R. Saunders. 1991. "Implications for the Release of Genetically Engineered Organisms," *J. Appl. Bacteriol. Symp. Suppl.*, 70:19S–30S.

Schechtman, M. G. 1992. "USDA Requirements for Safe Field Testing in the Environment," in *Genetic Interactions among Microorganisms in the Natural Environment,* E. M. H. Wellington and J. D. Van Elsas, eds., New York: Pergamon Press, pp. 289–297.